Monographs in Cool Climate Viticulture – 2

CLIMATE

David Jackson

Daphne Brasell Associates Ltd
with
Gypsum Press

First published in 2001 by Daphne Brasell Associates Ltd
PO Box 12–214, Thorndon, Wellington
Aotearoa New Zealand
http://www.brasell.co.nz
with Gypsum Press

ISBN: 0–909049–35–1

Acknowledgements
The author wishes to thank Paul Miskelly for valuable comments on this monograph; Graeme Buchan for his help, especially in Chapter 7; John Walsh of Hawkes Bay Windmachines and Neil Scott of Garden City Helicopters for information on the use of wind machines and helicopters for frost control; Damian Martin (Corbans Wines, Hawkes Bay) who kindly supplied data of bud burst and harvest dates for vines in several world situations and Gilbert Wells who produced the graphs in chapter 6. The author and publisher thank also, Federico Monsalve and Martin Prout for their management of the production of this book.

Edited by Federico Monsalve and Martin Prout,
Whitireia Publishing, Wellington
Designed by Hamish Thompson
Indexed by Martin Prout
Printed by Astra Print Ltd

10 9 8 7 6 5 4 3 2 1

CONTENTS

	INTRODUCTION	1
	What is a cool climate?	1
	Better wines, or different?	3
	Definition of a cool climate	5
1	WEATHER AND CLIMATES	6
	Tropical, sub-tropical and temperate climates	6
	Types of temperate climates	7
	Mediterranean	7
	Continental	8
	Climates with no dry season – maritime climates	9
2	CLIMATE – THE PARAMETERS	10
	Temperature	10
	Minimum temperature for growth	10
	Cold damage	11
	Heat damage	11
	Rain	13
	Wind	14
	Hail	15
	Coping with cold temperatures and other undesirable weather events	15
	Freeze injury	15
	Soil management for frost control and water conservation	16
	Wind breaks – advantages and disadvantages	17
	Frost reduction by heat application	18
	Frost reduction by mixing air	18
	Vine management	21

	Frost reduction by water sprinkling	22
	Macro, meso and microclimates	25
3	STYLE AND TERROIR	28
	Terroir	29
	Grande crus vineyards of the High Medoc	30
	The limestone soils of St Emilion	31
	The clay soils of Pomerol	31
	The climate underground	32
4	CLIMATE INDICES	34
	Factors affecting the speed of ripening	35
	Temperature effects	35
	Rainfall	35
	Continentality	35
	Altitude	35
	Latitude effects	35
	Some doubts	36
	Choosing an index	36
	Degree days (DD) also called Heat Units	36
	Mean temperature of the warmest month	38
	Latitude	38
	Length of the growing season	38
	Latitude Temperature Index	39
	The usefulness of climatic indices	40
	Seasonal comparisons	40
	Comparing districts	41
5	ESTABLISHING A VINEYARD – THE USE OF CLIMATIC UNDERSTANDING	43
	Preliminary investigation of the district	43
	Choosing an index	44

Getting the data 45
Choosing the grapes for the vineyard 45
Factors that will delay ripening and might make a site have lower ripening capacity than the index predicts (see also Table 4.1) 48

6 COOL CLIMATES OF THE WORLD 49
Southern England and Nova Scotia 49
Reims (Champagne, northern France), Christchurch (New Zealand) 52
Freiburg (Germany), Summerland (British Columbia, Canada) 54
Zurich (Switzerland), Corvallis (Willamette Valley, Oregon) 56
Dijon (Burgundy, France), Corvallis in the Willamette Valley (Oregon) and Christchurch (New Zealand) – the Pinot factor 58
Grapes in the cool/warm boundary of viticulture 60
Cabernet Sauvignon and other 'Claret' grapes 60

7 GLOBAL WARMING 63
Natural changes in climate: some historical consequences 63
Causes of climate change 64
Sun/earth cycles 64
Sunspot cycles 65
El Niño/La Niña 65
Cosmic collisions and volcanoes 65
Human input 66
The greenhouse effect 66
Global warming – some cautionary remarks 67
Consequences 68

REFERENCES AND FURTHER READING 70
INDEX 73

FIGURES

0.1 Flowers of apples and grapes 1

1.1 World map showing distribution of viticulture and the cold and warm ocean currents affecting this distribution 8

2.1 Frost and freeze injury

(a) Frost damage to young leaves and shoots 12

(b) Leaf variegation as a consequence of light frost damage 12

(c) Poor shoot growth following winter bud damage 12

(d) Crown gall infection on trunk following winter freeze damage 13

2.2 Botrytis – a common disease, especially difficult to control in wet climates 13

2.3 Hail damage to growing berries 16

2.4 Clean-cultivated vineyard in low rainfall non-irrigated Australian vineyard 17

2.5 Frost control

(a) Oil pots 19

(b) Oil pot – cleaner burning option 20

(c) Wind machine – propane driven 20

2.6 Effect of pruning on earliness of bud burst

(a) Early pruning 22

(b) Mid winter pruning 22

(c) Late winter pruning 23

(d) Pruning early spring 23

2.7 Frost protection using sprinklers

(a) Minisprinkler positioned on top of a post 24

(b) Position of minisprinkler between posts 24

(c) Susceptible tissue (apricot flowers) protected by ice from water applied by sprinklers 25

2.8 Mesoclimates in a hypothetical area 26

2.9 Percentage increase or decrease in radiation in north-or-south-facing slopes of 10°, 20° or 30° inclination 27

3.1 Factors contributing to style 29

3.2 Vineyard on south-facing slope in Champagne and Moselle 33

4.1 Heat distribution in two climates with similar Degree Days – Christchurch, New Zealand and Giesenheim, Rhine 39

5.1 Stevenson Screen and Pan Evaporimeter 44

6.1 – 6.13 Show data for representative climates in France, Germany, USA, Canada, Australia and New Zealand

6.1 Kew (UK) 51

6.2 Kentville (Nova Scotia) 51

6.3 Reims (Champagne district, France) 53

6.4 Christchurch (New Zealand) 53

6.5 Freiburg (Germany) 55

6.6 Summerland (British Columbia) 55

6.7 Zurich (Switzerland) 57

6.8 Corvallis (Willamette Valley, Oregon) 57

6.9 Dijon (Burgundy, France) 59

6.10 Bordeaux (France) 61

6.11 Napa (State Hospital, California, USA) 61

6.12 Napier (Hawkes Bay, New Zealand) 62

6.13 Mt Gambier (Australia) 62

7.1 Europe showing current boundaries for viticulture and expected northern boundary if global warming raises temperatures by 3°C 69

TABLES

4.1 Comparisons of effectiveness of Latitude Temperature Index (LTI) and Degree Days (DD) for evaluating suitable districts for viticulture 42

5.1 Grape varieties grouped according to ripening ability in different climates 46

INTRODUCTION

What is a Cool Climate?

In no other fruit crop does climate appear to play a more important role than it does for grapes. Apples, for example, are a temperate crop, although they mature better in areas where summers are not too cold and the climate not too wet. Being temperate fruits they do not perform well in sub-tropical or tropical climates. Different temperate climates produce higher or lower apple yields and quality may also be related to climate factors. Growers may be concerned with weather events such as

FIGURE 0.1 Flowers of apples and grapes. Left: Apples appear at the same time as leaves emerge in the spring

Right: Grape inflorescences appear well after bud burst; individual flowers on the inflorescence will not open until several weeks later

late spring frost, hail or sunburn. Most of these aspects also affect the grape grower but, in addition, he/she needs to consider that heavy rain near harvest will cause fruit splitting followed by disease, early autumn frost may cause leaf fall and even berry death, and short cool seasons will severely delay ripening.

Our special concern with climate and particularly heat accumulation is because the growing season for grapes is relatively short. For example, remaining with the apple comparison, apples flower in spring and the fruit has until autumn to grow and ripen. Thus they will grow in areas of Canada, England, Holland and Germany which are at a relatively high latitude and have a relatively short growing season. Grapes in the same areas often have to struggle to ripen, because, although grape buds burst at the same time as apples, grape flowers do not appear until mid-summer when considerable vegetative growth has occurred. Thus the grape must grow and mature in only half the growing season that is available to apples (Figure 0.1).

An explanation is needed as to why the topic of cool climates is special and worthy of a monograph series. Here are some features to consider:

1. Definition. A grape is growing in a cool climate when the mean temperature in the month before harvest is 15°C or below.

2. Grapes in cool climates have larger variation between vintages than those in warm climates.

3. In cool climates, poor vintages are mostly due to cooler-than-normal conditions resulting in lowered ripeness of the grapes. In warm climates cool seasons may result in later ripeness but not necessarily poorer quality.

4. Higher yields are normally achieved in warmer climates because buds are more fruitful. The vine in such climates has a greater capacity to ripen a higher crop. Yields in cooler climates are usually lower and if high yields are obtained the vine may not be able to ripen them to a satisfactory level.

5. Warm climates produce a higher level of sugar in the berries, which increases the alcohol level in wines, resulting in a wine with more body. Grapes in warm climates often ripen with low acid levels which may need to be supplemented

in the winery. Sometimes high pH levels in the wine cause wine-making problems.

6. Cool climates usually produce grapes with lower sugar levels than warm climates, even if the grapes attain full ripeness. In cool vintages it may be necessary to add sugar (a process known as chaptalisation). Acid levels may be high and in cool seasons they may have to be adjusted in the winery. The pH levels are seldom too high, which is usually considered to be a bonus.
7. Cool-climate wines although of lower body are often said to be more delicate and elegant. Higher acids give a sense of freshness not found in warm-climate wines.
8. Of the dozen or so world-recognised quality grapes there is a tendency to find more whites that are successful in cooler climates and more reds that are successful in warmer climates. Pinot noir is a notable exception – producing excellent 'Méthode Traditionelle' wines in cool climates such as Champagne, and fine Burgundy-style wines in slightly warmer but still mild climates. If a district is found to produce consistently fine wines from Cabernet Sauvignon or Shiraz (Petite Syrah) it is almost certainly a warm climate.
9. In cool climates it is common for viticulturists, in their selection of clones of specific varieties, to select an earliness characteristic. Rootstocks may be chosen for their ability to ripen the grapes before the norm.
10. Warm climates are more forgiving than cool climates and more skill is required by the cool-climate viticulturist to produce quality grapes.
11. Fortified wines are seldom produced in cool climates.

Better wines, or different?

In *The Production of Grapes in Cool Climates* (Jackson & Schuster, 1997) Danny Schuster and I were criticised for saying: 'a cool climate is one which will have the possibility to produce table wines of distinction'. As such it is not an incorrect statement except that it implies that grapes in warm climates cannot do this. They can and they do. Wines should be compared like with like. Those who prefer cool-climate wines, simply prefer cool-climate wines, they are expressing no superior judgement.

Nevertheless, the impression remains that many of the best table wines are produced in cool climates. Becker (1985) says:

> Under cool conditions white wines tend to be fresher, more acidic and of finer bouquet and aroma. In warmer zones the aroma forfeits its freshness and the wines have more alcohol and often lack balance. High quality white wines generally come from cooler zones whereas the optimum for red wines is somewhat warmer. In very warm regions wines are high in alcohol and short on taste and aroma; such zones are suitable for dessert wines.

Coombe (1987) suggests that this is too simplistic. According to him all we can accurately say is that ... 'in hot regions the increase in sugar and decrease in malate are more rapid, especially the latter. The evidence that hot regions give grapes that lack compounds contributing to fine flavour is indirect'. Rapid ripening means it is easier to miss the correct picking day for grapes, high temperatures at crushing and during fermentation may increase liability to oxidation faults. It could be argued that, traditionally, because more care has been lavished on vine management in cool climate districts the production of quality grapes is more likely to result.

Graham Due (1995) has argued that 'there is no causal relationship between climate and quality ... what is certain is that climate does affect the style of the wine: warm climates produce wines higher in alcohol and extract'. In principle the author agrees with this (style will be mentioned in more detail in Chapter 3).

We shall not consider the debate any further except to repeat that there will be differences in style between cool and warm regions. While it is pointless to argue which is better, we can be sure that management of vines and wines is different in the two situations and demands different approaches to viticulture and oenology.

Definition of a cool climate

We need, finally, to return to the earlier definition of a cool climate and elaborate the statement: 'a grape is growing in a cool climate when the mean temperature in the month before normal harvest is 15°C or below'.

The key to a cool climate is that grapes will be ripening in cool conditions which will impart special qualities to the resultant wines. In fact it is possible to have a district that has characteristics of both a cool and a warm climate. In Hawkes Bay on the east coast of the North Island of New Zealand, Müller Thurgau ripens early when the temperatures are above 15°C, Cabernet Sauvignon ripens late when the temperatures are below 15°C. To simplify such problems we have introduced two terms: *alpha zones* and *beta zones.* An *alpha zone* is an area where the specific grape in question ripens in temperatures in the last month before harvest of 15°C or below. It is a *beta zone* if the temperature is above 15°C. Thus Hawkes Bay is a *beta zone* for Müller Thurgau and an *alpha zone* for Cabernet Sauvignon. We will need to manage a vineyard with Cabernet Sauvignon with awareness of the special needs of a cool climate (warm mesoclimates, lower cropping, open canopies).

Despite these clarifications people will still wish to be able to designate a district's climate as cool or warm. We therefore define a cool climate as one where the majority of grapes grown in that district are growing in an *alpha zone*.

In our example of Hawkes Bay which grows many grapes that are later than Müller Thurgau but earlier than Cabernet Sauvignon the area can be designated as a warm climate, although in fact it is only marginally so.

CHAPTER 1

Weather and Climates

Weather is the daily variation in heat or cold, rain or snow, sunshine or cloud, wind or gale. Weather is a never-ending source of conversation and failing any other spicy news, broadcasters can always fill the gaps with a freak weather event which is certain to be occurring at some place in the world.

Climate is the sum total or average pattern of weather. It defines the expected weather patterns of a specific district, which may be wet or dry, windy or still, hot or cold, prone to thunder and hail or snow and sleet. In the rest of this chapter we will describe the parameters that contribute to a climate.

TROPICAL, SUB-TROPICAL AND TEMPERATE CLIMATES

Tropical climates are of course the warmest. They generally occur in latitudes from 25.5°N to 25.5°S and have a mean temperature of the coldest month above 18°C. Frosts are rare. Tropical plants are those which survive in these warm temperatures and they have very little, if any, tolerance to temperatures of 0°C or below. Tropical climates do not have significant temperature differences between summer and winter although many have rainy and dry seasons.

Grapes, which evolved in temperate regions, do not crop well in tropical climates. If they are to be grown, growers prune hard after harvest; this substitutes for the dormant (cold) period and the plant begins a new flush of shoot growth. If the vine in the tropics is spur pruned, shoots tend to lack flower bunches (*inflorescences*). If canes are used, usually only the end one or two buds grow but these can be fruitful. Grapes grown in tropical regions tend to be used for dessert purposes and wines made in such climates usually lack flavour and finesse. As a bonus tropical climates will allow grapes to crop two or more times a year.

Sub-tropical climates have a mean temperature of the coldest month (MTCM) of between 13°C and 18°C. They can be moist or dry often with wet and dry seasons. Slight frosts are sometimes encountered and some sub-tropical plants have tolerance to 0°C or slightly below. Winter is cooler than summer. Southern Queensland, much of South Africa, southern California and North Africa are sub-tropical areas. Sub-tropical fruits include pawpaw, pineapple, litchi, tamarillo and cherimoyas.

Because the sub-tropical winters are mild to warm, grape vines do not get clear-cut dormancy. Leaves hang on the vines well into winter and, afterwards, buds burst over a protracted period so that development of berries is not synchronised.

Buds are more fruitful than in tropical areas although generally less than in temperate areas. Fine wines are seldom produced in sub-tropical areas.

Temperate climates are of course the coolest of the three major groupings. Nevertheless, many such areas can have summer temperatures matching in heat those in tropical and sub-tropical areas – albeit for a limited period. Temperate climates have a distinctive temperature range and deciduous trees lose leaves over winter and the plant enters dormancy. Winter temperatures can be very cold and frosts can also occur in autumn and spring. The majority of the world's grapes and virtually all commercial wines are produced in temperate climates. Nevertheless, not all temperate climates produce grapes suitable for wine and thus it is important to categorise types of temperate climates.*

TYPES OF TEMPERATE CLIMATES

Mediterranean

Mediterranean areas are close to the sea but usually also close to a considerable land mass. These conditions occur in countries bounding the Mediterranean Sea. Summers tend to be hot and dry and winters cool or cold and damp. Some other areas in this classification are southern South Africa, southern Australia, parts of California and Central Chile. Grapes do well in these climates, the dry-hot summer helps ripening

*This is a simple classification and readers may be aware that other more detailed systems are available. This classification is perfectly adequate for our needs.

and reduces disease pressure and cold, or at least cool, winters reduce pest build-up. With irrigation, yields can be very high. Traditionally these areas have not been noted for fine table wines but some believe this will change.

Continental

These areas occur within a considerable land mass and are moderately far removed from the sea or lakes. They can range from cold to hot. Continental climates in temperate zones are often not ideally suited to grape production. For example, Figure 1.1 shows that most grape-growing areas occur between latitudes 30 to 50 which roughly equates to mean annual temperatures of 10° to 20°C. Some anomalies occur, thus in North America we find that the central areas produce few grapes, despite apparently suitable summer temperatures. The reason is the considerable continentality of these districts which means that, while summers are warm to hot and completely capable of ripening grapes, winters are very cold; so much that vines can be killed and viticulture becomes unsustainable.

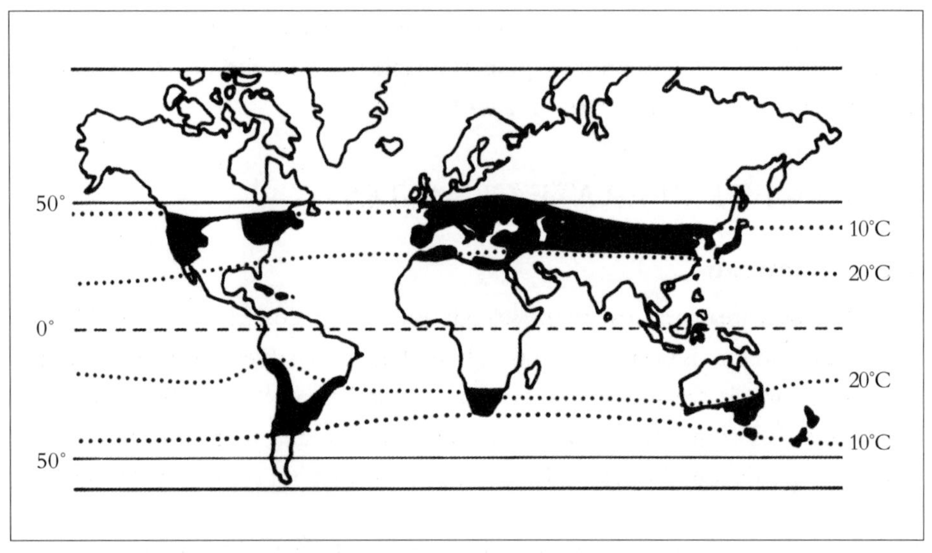

FIGURE 1.1 World map showing distribution of viticulture and the cold and warm ocean currents affecting this distribution (After Johnson, 1985; Coombe, 1987; Fullard and Darby, 1979)

Continentality is the term used for the difference between the mean temperatures of the warmest and coldest month. As we move east in Europe we find that continentality increases from northern France, to Germany, to Poland, to Russia. Some figures (°C) for other world areas are:

1. *Coastal – maritime*: San Francisco (USA), 6.7; Hobart (AU), 8.7; Blenheim (NZ), 10.3; Kew (UK), 13.4; Bordeaux (FR), 13.8.
2. *Inland but with some maritime influence*: Reims (FR), 15.5; Portland (USA), 15.7; Dijon (FR), 17.8.
3. *Inland – continental*: Prosser (Eastern, USA), 22.6; Graz (Austria), 22.8; London (Ontario, Canada), 26.4.

Climates with no dry season – maritime climates

Most traditional and quality wine regions occur in such climates. They include northern France, Germany, Switzerland and northern Italy. The New World areas are: Oregon, eastern USA, Tasmania and New Zealand. In many of these, droughts can occur in summer when evapotranspiration exceeds natural rainfall. Irrigation is used in areas where plant stress is severe and regular.

Variation in summer rainfall can affect suitability for grape production. In the Rhine and Moselle regions of Germany summers can be moderately wet but in autumn rainfall is low which reduces disease incidence. In the Willamette Valley in Oregon the summers are dry but spring and autumn are wet – making disease control more difficult. The eastern USA has never had the profile of California for grape production despite adequate heat, this is due to the higher summer rainfall which makes disease control more difficult. Many eastern areas have very cold winters. In a later chapter (Chapter 6) we examine more closely several world climates.

Our survey does not cover all world climates. We have used a few examples where, in fact, many exist. Asian areas have not been mentioned even though grape production there is significant. So far, however, in these areas only a very small proportion of grape production has been used for wine. Areas producing fortified wines and brandy will not be discussed because the emphasis in this series of monographs is on production of table wines which is the dominant use for grapes in 'cool' climates.

CHAPTER 2

Climate – the Parameters

In this chapter we shall look at the physical factors that make up a climate and their interaction with the vine's biology.

TEMPERATURE

Temperature is the key factor and the one most often discussed in relation to viticulture; its importance is related to the following features.

Minimum temperature for growth

Below 10°C grapes make little growth and so vine development does not begin in spring until that temperature has been reached. Likewise in autumn, growth and/or development ceases as the temperature drops to that level.

As the temperature rises above 10°C, growth of all parts of the plant increases until an optimum rate of growth is reached at between 25 and 32°C. Above these temperatures growth declines and eventually falls to zero. High temperatures can kill the plant especially if water availability is low. (Loss of water through the leaves cools the vine when water is adequate.)

The more heat the vine receives the earlier will the grapes mature. In districts where the season's accumulated heat is too low, grape production is not feasible because the fruit will not mature before autumn frosts and cold weather stop growth; in warm climates the grape ripens well before the autumn.

Temperature has some other effects on vines. Sugar, as noted, increases more rapidly and attains higher levels in warm climates. Acids drop with maturity, but the two main acids have different characteristics: tartrates drop with increased maturity but the final level will be little different in cool or warm climates. Malates on the other hand will drop more rapidly in warm conditions. Thus cool climates have malate dominance which may give wines an apple flavour. (Malic acid is the dominant acid in apples.)

Cold damage

Vines are damaged at temperatures below –15°C. Some varieties such as Cabernet Sauvignon, Sauvignon blanc, Sémillon, Chenin blanc and Grenache are more susceptible than others, eg Riesling, and Pinot noir. Many grapes of American origin and Franco-American hybrids are tolerant of temperatures below –26°C. Extremes of tolerance vary from *Vitis amurensis* –30°C, to Grenache –14°C.

It has been suggested, in the European context, that: in a climate where the mean temperature of the coldest month is below –1°C, vines are likely to be susceptible to winter freeze, or if the winter temperature falls below –20°C in more than one year in 20 then the profitability of viticulture is doubtful because of heavy winter vine damage (Becker, 1985).

As buds swell in the spring their susceptibility to cold damage increases. Thus a bud which in the middle of winter can survive –15°C will lower its resistance to –1 or –2°C at bud burst. From that point onwards buds, leaves, and fruit are damaged by temperatures below about –1°C (Figure 2.1). Frost in the spring kills emerging buds and their shoots, as well as any developing fruit. New growth will occur, but these secondary shoots are much less fruitful than the primary ones and, in addition, their coming later in the season, will delay maturity. In autumn, early winter frosts will kill leaves and the fruits remaining will have to complete their development without the carbohydrates contributed by those leaves. A more serious frost will severely damage the fruit. As can be seen, frosts in susceptible areas can be a major problem for successful viticulture. Frost control will be discussed later in this chapter.

Heat damage

As temperatures rise during the growing season berries become susceptible to sunburn. Most susceptible are berries that have developed in shaded regions of the vine and which, for one reason or another, later become exposed to direct sunlight.

FIGURE 2.1 Frost and freeze injury.
Photographs courtesy of Glen Creasy
(a) Frost damage to young leaves and shoots

(b) Leaf variegation as a consequence of light frost damage

(c) Poor shoot growth following winter bud damage

(d) Crown gall infection on trunk following winter freeze damage

FIGURE 2.2 Botrytis – a common disease, especially difficult to control in wet climates

Berries exposed from their early development are more resistant but even these can be damaged as temperatures rise above 32°C. Damage can be in the form of a slight sunburn which does not kill any part of the berry. Such fruit may contain slightly greater quantity of tannins, which, if widespread, can impair the fruit quality. More serious damage may kill the exposed berry parts or even the whole berry. Sometimes half the bunch can be destroyed. Generally the damage looks worse than it is.

RAIN

Rain at the right time and at appropriate intensity is beneficial. At the wrong time it can be, to say the least, unwelcome. Negative aspects are outlined below.

1. Heavy rain close to harvest can cause the berries to split due to excessive uptake of moisture. Split berries are very susceptible to diseases – especially

those which reduce wine quality – eg botrytis fungus (Figure 2.2) and acetic acid bacteria.

2. Rain, either directly, and/or due to the associated high humidity, can make conditions ideal for many fungal disease even if splitting does not occur. Rain will also tend to reduce the rate of ripening even though temperatures remain warm. A district with high rainfall may ripen its grapes later than another district with the same accumulated heat during the season. Rain also makes vineyard operations more difficult and at times even impossible. Thus heavy rain can prevent the application of sprays at the most appropriate time or may delay harvest past the time when the grapes are at their optimum maturity.

   Note: *Botrytis cinerea*, induced by moist conditions at véraison,* may sometimes have positive effects on quality. This occurs if after rain, cool dry conditions prevail. Berries shrivel but do not split and sweet 'botrytised' wines may be produced.
   Some districts will produce these conditions occasionally – eg Rhine and Moselle in Germany. In other areas, eg Sauternes, they occur more regularly.

WIND

Some areas are prone to winds during the growing season. Strong winds can cause physical damage in a vineyard. Before lignification (the development of woodiness in the stem) shoots may break at the base and the shoot and its fruit will be lost. Moderate wind may cause the young shoots to move against the supporting wires and break off the shoot tip – the shoot then forms branches and becomes difficult to train.

The microclimate of a vineyard can be altered by wind. In calm conditions the air around the vines heats up in the sun and the temperatures may be 10°C warmer than a similar situation exposed to a cold wind. In German vineyards, winds of more than 2 m/sec along the rows and 4 m/sec across the rows have been shown to destroy this special microclimate. Wind promotes transpiration and therefore water loss is greater thus slowing down the rate of plant growth in dry conditions.

* The stage of berry development when berries start to soften and change colour, initiating the accumulation of sugar and respiration of acids.

While winds are generally unwelcome, there are positive aspects that need to be mentioned. Winds increase drying after rain and lower the humidity thus reducing the possibility of diseases. Wind also dries the vineyard floor and enhances the viticulturist's ability to use machinery after rain. It will move foliage in a vine canopy and will allow flecks of light to reach otherwise shaded leaves or fruit. Wind can reduce the incidence of frost by mixing warmer air above with cold air at vine level.

HAIL

Hail will damage leaves and fruit of vines. Mostly the damage is not too severe and the result is a minor proportion of lost berries (see Figure 2.3). Occasionally the major part or the whole crop can be damaged and severe physical damage to the canes can be sustained. For example, in the 'hail belts' of eastern Australia whole crops of particular vineyards in the erratic paths of summer thunderstorms are regularly lost. The Hunter Valley and Orange in New South Wales are especially prone.

COPING WITH COLD TEMPERATURES AND OTHER UNDESIRABLE WEATHER EVENTS

Freeze injury

As noted different species and varieties vary in their cold hardiness. Clearly in areas with severe winter freeze, cold resistant varieties are recommended. Other ways to reduce damage are:

1. Avoid excessive vigour since vines which grow late into the season are slower to acclimate to cold.
2. Consider the use of multiple trunks so that if one is killed another may survive.
3. Mound the base of the trunk with soil in winter, this may reduce injury to the lower trunk and the graft union.
4. In extreme cold the whole vine could be removed from the trellis and buried under soil.

FIGURE 2.3 Hail damage to growing berries

Soil management for frost control and water conservation

Bare soils, which are firm and preferably moist, can increase temperatures in the vineyard over and above ones with weeds, grass cover, or loosely cultivated soil. This is because such soils absorb more heat during the day and re-radiate it at night. Even during the day, temperatures are warmer under such management. The total accumulated heat is greater with bare soil and ripening is advanced, a result welcomed by growers in cool climates. Bare soil radiating heat at night will raise the temperature in the vineyard by 1 to 2°C, which in many cases is sufficient to prevent frost damage. If grass is grown between rows of vines it is wise to keep it cut very short, especially in periods of possible frost damage.

A bare soil loses less moisture than one with weeds or cover crops. In dry areas where irrigation is not available water conservation is very important (see Figure 2.4).

It must be noted that on steep slopes the advantages of heat accumulation may be offset by susceptibility of bare soil to erosion.

Wind breaks – advantages and disadvantages

Wind breaks (shelter-belts), which can be a growing line of trees or synthetic netting constructed around the perimeter of the vineyard, are beneficial in many situations. They will reduce the disadvantages of wind as listed (reduce physical damage, reduce water loss and increase total accumulated heat). They do, however, reduce space available in the vineyard; living hedges will poach water and nutrients from the neighbouring vine rows and shelter may reduce quick drying in the vineyard. If wind breaks stop air drainage in the night, frosts may be increased. Sometimes growers will plant trees in the first few years but remove them later when the mature vines provide a degree of self-sheltering. As a guide it can be assumed that

FIGURE 2.4 Clean-cultivated vineyard in low rainfall non-irrigated Australian vineyard. Water loss is reduced by the low ratio: leaf area to soil volume – ie less evapotranspiration

reasonable protection will be obtained for a distance of 10-times the height of the shelter on the leeward side. Living shelter-belts provide nesting and perching sites for birds and usually increase problems of bird damage.

Frost reduction by heat application

It is not common to heat a vineyard growing wine grapes, although greenhouse production has long been used for dessert grapes. Some viticulturists have tried plastic covers over the vines to reduce rain and give localised heating but this has not been done at more than an experimental level.

For frost control heat application has been used in various areas. Normally diesel is burnt in 4–5 litre pots distributed through the vineyard; the diesel being ignited when the temperature reaches 0°C. Up to 100–120 heaters per hectare may be positioned in a vineyard. When the temperature falls to +0.5°C about 30 per cent are lit. The grower then monitors the temperature in the coolest part of the vineyard and lights more heaters when the temperature then falls to below zero. Heaters are stopped when the temperature returns to 1°C. This method, which is quite effective, is not popular with neighbours and those concerned with the environment since it produces copious quantities of black smoke. 'Smokeless' pots (Figure 2.5) can be purchased which are much more environmentally friendly, but expensive. Igniting the burners in the middle of cold nights can also be an inconvenience to the growers.

Frost reduction by mixing air

Frosts in spring or autumn are normally radiation frosts. Here cooling is due to radiation of heat from the vineyard to the sky and this occurs during still cloudless nights when the humidity is low. Cold air, being denser than warm, sinks to the lower parts of the vineyard and when the temperature drops to dangerous levels damage occurs. Over the course of the night the thickness of the layer of cold air increases. The boundary between the damaging and non-damaging air is called the inversion layer (IL). The temperature difference between 1.0 m and 5.5 m is a convenient measure of the 'inversion strength'.

FIGURE 2.5 Frost control. (a) Oil pots: top, lighting torch contains 50:50 petrol:diesel mixture. There should be a metal gauze between spout and reservoir. Below, 5-litre oil pots; triangular 'spider' on top can be added to reduce burning rate

(b) Oil pot – cleaner burning option

(c) Wind machine

One way to overcome frost damage is to mix the air and so bring the warmer air down into the vicinity of the vines. If the inversion strength is 6°C, the mixing of air can lift the temperature at 1 m by 2°C. Inversion strengths below 4.5°C may be insufficient for frost control.

Air mixing can be done in two ways: helicopters and wind machines. The effectiveness of helicopters depends on the size of the machine. A small helicopter (625 kg) will lift the temperature over 10 ha by 2– 4°C if the IL is 30 m or below. A larger machine (say 2000 kg) will cover 50–70 ha at the same IL height. Such a machine will be needed for ILs up to 60 m – the area covered in such circumstances will be between 10 and 50 ha.

Specially-mounted fans or 'wind machines' can be placed at strategic positions in the vineyard. Growers will need to check with local producers or distributors concerning precise details of the machine's capacity, but the following will provide a general guide. One fan mounted on a tower of 10 to 10.5 m height and with 5.4 m diameter blades and 80–100 kw power output will cover 4 to 6 ha, raising the temperature 2 to 4°C with ILs of up to 50–60 m. Machines can be powered by electricity, diesel, petrol, or LPG. Even when there are cold conditions but little or no inversion, wind machines can be used with diesel-fuelled orchard heaters to provide reasonable protection. Growers will need to gain local knowledge of frosts, particularly the expected height of the IL.

Fans are turned on or helicopters sent aloft when the temperature drops to 1°C.

Vine management

Another approach to take advantage of the temperature gradient above the ground is to raise the base wire to the top of the trellis. This will mean adopting a curtain system of training, eg the Geneva Double Curtain or the Single Curtain. The temperature at 2 m may be 2°C above that at 30 cm.

The time of pruning can be used as a method of lowering frost risk. The principle is to delay bud burst to a time when the frost risk is lower.

The closer the time of pruning is to bud burst the later will be that bud burst – up to two weeks (Figure 2.6). The problem with this is that the considerable work involved in pruning will have to be completed in a very short space of time in late winter – early spring.

Growers will sometimes select the correct cane or spur number but leave canes unshortened. At bud burst, canes are pruned back either to short canes for cane pruning or spurs for spur pruning. These basal buds are slower in developing than the terminal ones, and being later they are slightly less susceptible to frosts at that time. (The End-Point Principle – see monograph No 1 in this series.)

Frost reduction by water sprinkling

If water is sprayed over a vineyard when the temperature is below 0°C, a layer of water will freeze on the outside of the tissue (Figure 2.7). Surprisingly this can be used as a method to reduce damage. The physics is as follows: As water freezes it

FIGURE 2.6 Effect of pruning on earliness of bud burst. Photographs courtesy of Adam Friend

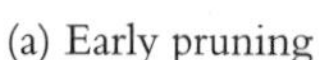

(a) Early pruning

(b) Mid winter pruning

releases heat – termed 'latent heat' – this heat prevents the temperature of the ice from dropping below 0°C. Fortunately, vines are able to tolerate 0°C so, as long as there is water in the process of freezing, the plant tissue will not fall below 0°C. The key to success is to apply sufficient water to ensure an appropriate amount of latent heat is continuously released. A sprinkling rate of 2.5–3.0 mm per hour will normally be adequate for an average frost. The sprinkling should begin at 0°C and terminate at 0 to 1°C.

The method is clean and effective. Heavy applications can cause problems with certain soils and a relatively high water use is demanding of pump capacity and water availability. There is also some breakage of young shoots due to the weight of ice. The simplest way is to have alkathene tubing along the top of the fence structure with micro-jets projecting the spray along the row (Figure 2.7). Irrigation engineers will usually be required to plan such a system.

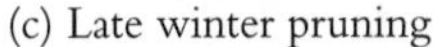

(c) Late winter pruning

(d) Pruning early spring

FIGURE 2.7 Frost protection using sprinklers

a) Minisprinkler positioned on top of a post

(b) Position of minisprinkler between posts

(c) Susceptible tissue (apricot flowers) protected by ice from water applied by sprinklers

Sometimes a frost can be a result of cool air, perhaps off a mountain, passing over and through the vineyard. Such 'advective' frosts are more reliably controlled by sprinklers than heaters or wind.

MACRO, MESO AND MICROCLIMATES

A ***macroclimate*** is the dominant climate in a designated geographical area, however, not all parts of that area will have identical climates. Where variation occurs we can subdivide that area into mesoclimates. A hypothetical situation is shown in Figure 2.8. The climates vary according to altitude, slopes, hollows, prevailing wind or its absence, aspect (slope of land in relation to the sun) and even manmade structures such as planted trees and shelter-belts.

These ***mesoclimates*** (often incorrectly called microclimates) are very important in cool climates because they vary the heat accumulation and frost risk and modify the range of cultivars that can be grown. For example, the best slopes of the Rhine and Moselle are used for Riesling which would not ripen adequately on lower slopes and flats with less-favourable aspects. In these lesser positions, earlier-ripening grapes such as Müller Thurgau and Sylvaner are used.

Figure 2.9 (from Becker, 1985) shows the increase or decrease in heat absorbing capacity according to the nature of the slope and whether it faces north or south Degree-day accumulation can be 50 per cent higher over the ripening period in a sun-facing slope compared with vines grown on a plateau.

Microclimates in viticulture refer to the immediate surroundings of the vine. They can be modified by growers who may change training systems to allow more sun to be intercepted by the leaves, pluck leaves from the vine near the berries for better fruit exposure or adopt grass or bare soil beneath the vines to increase or decrease ambient temperatures.

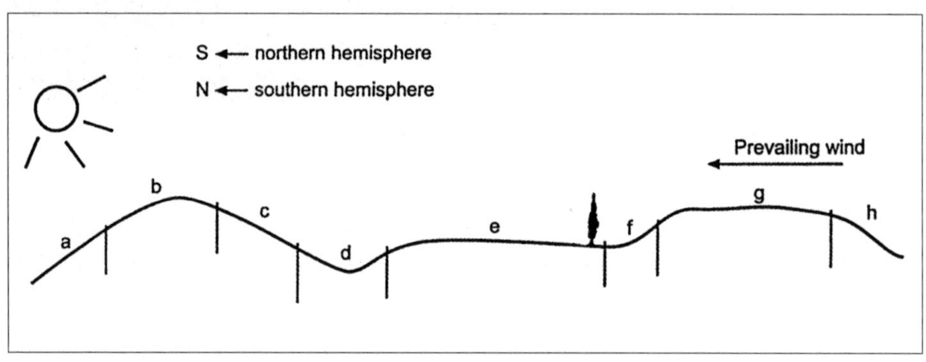

FIGURE 2.8 Mesoclimates in a hypothetical area. (a) A warm site catching more sun owing to the lie of the land. It misses late spring and early autumn frosts, since the cold air will drain to low-lying areas; there is also shelter from the prevailing wind. (b) The advantages of (a) will be counteracted by the cold which comes with increased altitude. (c) A cold site; although it may miss frosts in spring and autumn, it will accumulate much less heat in summer due to exposure to wind and a poor angle to the sun. (d) A cold site – very susceptible to frost; cold air from surrounding districts will drain into this area. (e) Still frosty but less so than (d). Some shelter from the wind may be obtained from the windbreak. (f) The windbreak, densely planted at the base of the hill, prevents cold air from draining away and a potentially frost-free site has been lost. This area would be shaded by the windbreak. (g) Less frost than (e), but a prevailing cold wind may slow the accumulation of heat units in summer. (h) Cold, like (c) above (Jackson and Looney, 1999).

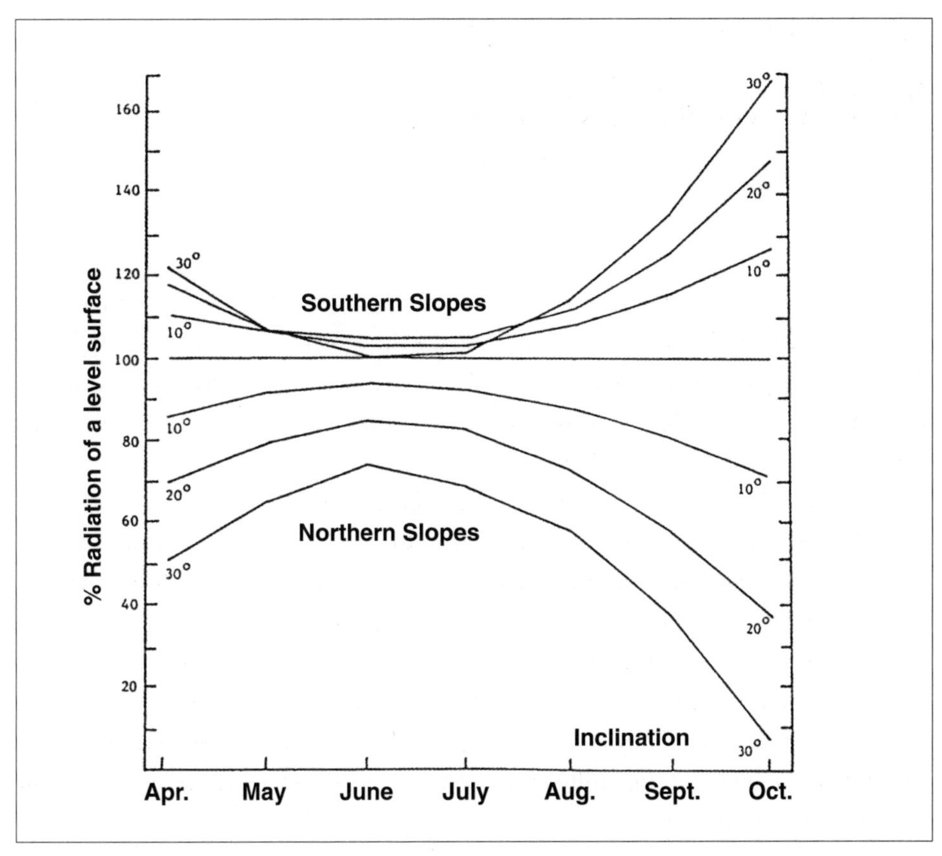

FIGURE 2.9 Percentage increase or decrease in radiation in north-or-south-facing slopes of 10°, 20° or 30° inclination (data after Becker 1985, in Rhine valley, 48° 15', northern hemisphere)

CHAPTER 3

Style and Terroir

On my last visit to French vineyards a statement that I heard more than once was: 'We don't want, necessarily, to make the best wine in the world, what we do want to do is to make a wine that is distinctive and characteristic of the area. A wine that people will remember and will return to, knowing that its chief characteristics will always be there'.

To me this is 'style'. Something that is characteristic of an area or a wine-producer, necessarily good but not necessarily the best, variable from place to place or time to time, but only within limits, and, most of all, reliable. By going for style we are going for the known against the unknown.

Style is initially a result of three primary factors: the *climate*, the *soil* and the *grape*. These determine the range of characters – the chemicals – that will be dominant in wine from that area. They are, however, themselves subject to secondary factors, which may modify and hopefully enhance the style. These are more subject to the viticulturist's control and are:

1. The creation of meso and microclimates by soil and plant management practices.
2. The modification of soil features by addition of fertilisers plus surface management by tillage mulching or use of green cover.
3. Selection of clones of the chosen variety.

The winemakers add their touch to the style.

Figure 3.1 summarises these points.

FIGURE 3.1 Factors contributing to style

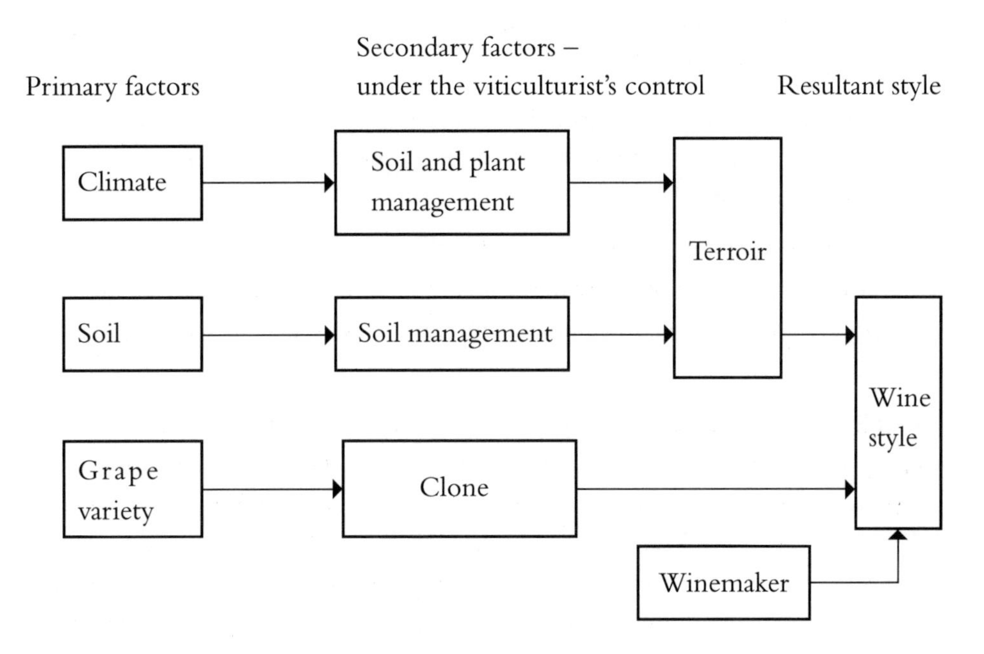

TERROIR

Terroir has no English equivalent, which accounts for the common retention of the French word in English. It refers to the immediate surroundings of the vine that ultimately put their imprint on the grapes and the wine produced from those grapes, creating style. Those surrounds include the fertility of the soil, its texture and structure, plus its depth and penetrability to roots. Terroir includes the soil's capacity to hold water and its ability to absorb or shed water after heavy rain. The soil can be heavy or light and can rapidly or slowly heat up in the spring. Climate interacts with soils and their management, slope and aspect to create meso and microclimates. Pruning and training methods may also modify microclimates.

One approach to understanding the concept of terroir is by outlining an *ideal* seasonal pattern of growth which might be as follows:

1. In early spring adequate water, nutrition and temperature allow buds to burst and shoots to grow evenly with sufficient vigour to attain about 17 nodes by the time of flowering.
2. Close to the onset of flowering the vine encounters a constraint – causing the shoot to terminate growth. This constraint may be an inadequacy of water, nutrition or both.
3. Post-flowering conditions are such that the vine remains in a healthy state but the constraints on shoot growth remain. In other words the stresses are sufficient to curb shoot elongation but not enough to debilitate the development of fruit.

This ideal is seldom exactly met; often the constraint will come late and some shoot tipping and trimming will be required; at other times the stress will be insufficient or too severe. Essentially a good terroir is one where variation from the above ideal is not excessive; a poor terroir is one where these conditions are seldom, if ever, achieved and where, to produce good grapes, a considerable input from the viticulturist is required. Because of this the terms '*holistic*' or '*integrated*' have been used to describe viticulture in good terroirs, whereas '*interventionalist*' is used for other sites (D Martin, pers. comm.).

It will be instructive to see how good terroirs are created by a beneficial interplay between soil and climate in specific areas. Three examples of recognised terroirs in the Bordeaux districts are presented. Data comes from Seguin (1986).

1. Grande crus vineyards of the High Medoc

These soils have a free water table within reach of the roots at the beginning of the season. The water table lowers progressively from spring to the beginning of autumn. By August the vine has utilised the freely available water and the shoots stop growing. Older vines have deep roots that keep the plant healthy but young plants may suffer excessive stress. If heavy rain falls prior to harvest the very penetrable soils allow good drainage which minimises the chances of berry splitting and other problems due to excessive rain and humidity.

2. The limestone soils of St Emilion

Here the soils are very shallow and the roots do not penetrate more than 70 cm. Potentially these soils are in danger of severe water stress. However, while the soil's relatively rapid use of water inhibits vine vigour, the nature of the limestone is such that capillary action brings up sufficient water to the roots to ensure that the danger of severe drought is minimised.

3. The clay soils of Pomerol

These soils have considerable clay content which can be as high as 60 per cent and in these asphyxiating conditions many young roots may die each year. Root penetration is seldom below one metre. Clay soils do hold considerable moisture, but the slow movement of moisture in such soils ensures devigoration of the vines. Interestingly, the fungal problems associated with heavy rain between véraison and harvest are ameliorated by the nature of the clay. A clay soil initially absorbs water but, as the clay swells, water penetration is reduced, thus lowering the danger of berry splitting.

Some comments are needed on these three soils:

1. All three soil types have some factor(s) which limit shoot growth, especially later in the season, but this is not excessive and does not interfere with fruit development.

2. No one type of soil seems to be essential.

3. Nutrition of the vine is not highlighted in any instance and in fact commentators have difficulty in relating quality with any specific nutrient. Nevertheless, shortage of one or more elements can be one factor in keeping a hold on vigour and, by and large, few, if any, good terroirs are found to be excessively fertile.

4. All three soils have mechanisms that reduce the effects of an influx of water after heavy rain. Not noted above, though common, is the use of slopes not only to assist heat accumulation but also to aid rapid drainage after heavy rain.

5. Also not noted above is the use by viticulturists of appropriate management practices to complement the features of the terroir. These range from the types of trellis and pruning used to preferences such as small vines and close planting of vines.

THE CLIMATE UNDERGROUND

As we have seen, the climate influences the expression of a terroir. For example, excessive rain will limit the attainment of a good terroir, excessive heat will put stress on a plant – a factor that may be difficult to control. Mostly when we talk of climate we think of the conditions above the ground – typically as measured in the standard Stevenson's Screen, one metre above the ground. Nevertheless, conditions below the soil surface are also important. Rain and its effects on water availability to roots has been discussed in this chapter but soil temperature is also important.

To some extent, roots and plant tops respond separately to temperature. A good example is the striking of cuttings in a hot bed. To give a cutting a good start the base of the cutting can be placed in a sand or soil mix at (say) 20°C while the tops are maintained at a cooler temperature (say) 10°C. Roots grow quickly from the base while the buds above remain apparently dormant. This holds true for any plant, including grapes, which root from dormant cuttings. Nevertheless, while in general this is also true for plants in the field, there is evidence that vines which grow in sites where the soil warms up quickly in the spring have a growth advantage above the ground. This advantage is carried through to harvest. Good terroirs often display early warming and of course this is of special value in cool climates.

Heavy moist clay soils take the longest to warm up; light and especially stony soils are the quickest. Soils on sun-facing slopes warm up more rapidly and, as noted earlier, bare soil surfaces show advantages over grass or weedy soil surfaces (Figure 3.2).

(a)

(b)

FIGURE 3.2 Vineyard on south-facing slope in Champagne ((a) above) and Moselle ((b) below.)

These slopes absorb more light and heat, reducing frost risk and aiding fruit ripening. The slope reduces rapid influx of water after heavy rain, thus reducing splitting and fungal infection

CHAPTER 4
Climate Indices

Climate indices have been introduced in Chapter 1 where one such index – Heat Units (Degree Days) – was briefly described. We now explore indices in greater depth.

An index has two main uses. First, it can tell us the seasonal variation in a specific district; thus by comparing the present season with the long-term average, the index can also help monitor whether the temperature has been warmer or cooler than normal. We may also monitor the heat accumulation as the season progresses and this may indicate whether the harvest is likely to arrive early or late. As will be seen later, some indices do this better than others.

The second value of an index is in comparing the potential of a new district for viticulture and/or determining which varieties are the most likely to succeed in such an area. Clearly this has more use for those who are still exploring the potential of their areas for viticulture. This is more common in the New World than the Old.

We shall see that indices can sometimes be misleading and provide false information. Nevertheless, providing we are aware of their limitations indices are most certainly worthy of study.

We shall therefore consider the factors that influence ripening and which might be used to create a suitable index. Then we look at some indices and consider their value to the grower. Finally, we examine the speed of ripening of various grape varieties which, by relating the data per variety to the index, tells us which varieties should be successful in a specific district.

FACTORS AFFECTING THE SPEED OF RIPENING

Temperature effects

Warm temperatures speed up the rate of ripening and most indices include a temperature factor. Increases in temperatures do not speed ripening at the same rate – an increase from 10 to 20°C advances growth processes more than does an increase from 25 to 35°C for instance. This is an important factor to consider when investigating indices.

Rainfall

Theoretically, rainfall should not be appropriate as a ripening index. In a rainy season ripening can be delayed but this would seem to be due to the fact that weather is normally cool in rainy conditions. Nevertheless it has been suggested that, even though an index contains a temperature measurement, as in degree days, an area with high rainfall will sometimes have less capacity to ripen specific varieties than the index suggests it should (see Jackson and Cherry, 1988, for evidence on this aspect).

Continentality

We have seen in Chapter 1 that areas which have continental climates often have such cold winters that grapes do not easily survive. Although indexes may show there is sufficient heat for ripening it may be impractical to grow vines because of cold temperature damage and vine death.

Altitude

Altitude affects the temperature of a district. Nevertheless an altitude factor is seldom added to climate indices since its effect should be mirrored in the temperature measurement. There are, however, situations where an altitude factor may be needed – see below under Latitude Temperature Index (LTI).

Latitude effects

Latitude has, in general, a negative relationship with ripening capacity. Figure 0.1 indicates that most wine grapes are grown in areas between latitudes 30 to 50 and of course the lower the figure the higher the ability to ripen grapes. Thus latitude alone could be used as a ripening index, a possibility which will be discussed later.

In high latitudes, mid-summer days are long and it is reasonable to assume that extra hours at mild or warm temperatures will contribute more photosynthates* than the same temperature at lower latitudes where days are shorter. Huglin (1978) and Gladstones (1992) suggest adding adjustments to the calculated degree days to cope with this factor.

Some doubts

Some Australian workers (eg Due *et al*, 1993) have correlated weather events in various Australian vineyards with ripeness and other factors. Their data suggests that weather measured in meteorological screens correlates with date of bud burst but with little else. Due (1994) suggests that factors such as crop load, site management and pruning and training can be at least as significant. Our data show that climate contributing to a specific index is important, but we would not argue that other factors can have significant effects and, as mentioned below, these must be taken into account by the intelligent viticulturist.

CHOOSING AN INDEX

We shall now list a number of climatic indices based on the parameters just discussed and indicate their usefulness for viticulture.

Degree Days (DD) also called Heat Units

Because this is the best known and most used, we have already commented on some of its features.

Degree Days (°C) are usually calculated using the formula:

DD = (mean monthly temperature – 10) x number of days in the month.

The annual DD is the total for the months where the temperature is above 10°C.

It is not difficult to appreciate that in a month where the mean temperature is

* *Photosynthesis* – the process whereby light falling on leaves activates the conversion of water and carbon dioxide to sugars, used for energy and growth. Sugars and other materials are collectively called *photosynthates*.

10°C or less, DDs will be recorded as zero, whereas in fact there may be days when the temperature will be above 10°C and some growth may be expected on those specific days. Degree Days can also be calculated on a daily basis; here the annual DD is the sum of all those days where the temperature is above 10°C. This calculation probably therefore gives a more accurate figure although in fact it is not often used. The annual DD will be higher if daily calculations are used. Wendland (1981) has provided a formula which will enable the user to calculate with reasonable accuracy the conversion of one to another. The key thing to remember is that it must be specified which method of calculation is used and to make any comparisons using the same calculation method. Monthly calculations are used in this book.

Because we may be dealing with some districts which are very hot and where the winter is mild, heat can still be accumulated in autumn and early spring when plants may still be dormant. Viticulturists often use just seven months, ie April to October – northern hemisphere; October to April – southern hemisphere. Thus they leave out those months which may accumulate DDs but where active vine growth is not occurring. Once again when making comparisons it is important to specify whether all months are used or just the seven summer months. This book uses all months above 10°C.

Amerine and Winkler (1944) use the following DD classification:

Region	I	Cool – below 1390 Degree Days
	II	1390 – 1667
	III	1667 – 1945
	IV	1945 – 2220
	V	Very hot – above 2220 Degree Days

These figures provide a guide to the varieties that might be grown in each region. Most high quality wines are produced in areas that have temperatures in the range of groups I and II.

The author believes that for comparing the ripening capacity of a district the use of DDs is not always sufficiently accurate and this is especially so when making comparisons between cool climate districts. For this reason another index is favoured – see the Latitude Temperature Index.

Working in the Australian situation, Gladstones (1992) has made a number of refinements to the DD calculations which he maintains give more accurate indicators for new viticultural districts. One sensible suggestion is that because the difference in response to temperature is greater below 20°C it is more accurate not to count any temperature increase above 19°C. The reader wishing to delve into this subject in more depth may like to refer to a debate between Gladstones and myself listed in the references at the end of the book.

Degree Days is the main index used to compare different seasons or compare variations within macroclimates, ie mesoclimates.

Mean temperature of the warmest month

In their two books: *Viticulture Vols I* and *II*, Coombe and Dry (1988; 1989) simply use the mean temperature of the warmest month (MTWM) (normally January in the southern hemisphere; July in the northern hemisphere) to indicate climatic zones. Within Australia it generally works well and does not require complex calculations or the searching out of less accessible data. It does not necessarily translate directly to other areas when comparing different climatic zones.
MTWM has little value for seasonal climatic comparisons.

Latitude

Latitude has, generally, an inverse relationship with ripening capacity, and could be used as a temperature index, however, on the world scale it can confuse. For instance at 51° latitude in England, early-ripening grapes will mature; yet they will not mature at a similar latitude in northern America. The reason is that the Gulf Stream warms much of western Europe while the massive area of cold tundra in northern Canada shortens the season in northern America. Latitude is not generally used as a climate index but, as we shall see shortly, it has been used as part of another index.

Length of the growing season

As a generalisation it is often said that the minimum length of the growing season for grapes to ripen is 182 days. Length of the growing season is seldom used as an index although it could be calculated from weather data as the number of days where the mean temperature is above 10°C.

Latitude Temperature Index

This Latitude Temperature Index (LTI) has been developed at Lincoln University, New Zealand, using two previously described parameters: (MTWM and latitude).

$$\text{LTI} = \text{MTWM} \times (60 - \text{latitude})$$

Figure 4.1 shows graphs of two grape-growing districts – Geisenheim on the Rhine in Germany and Christchurch, South Island, New Zealand. Both districts have almost the same DDs – 970 – but their ripening ability is different. Geisenheim

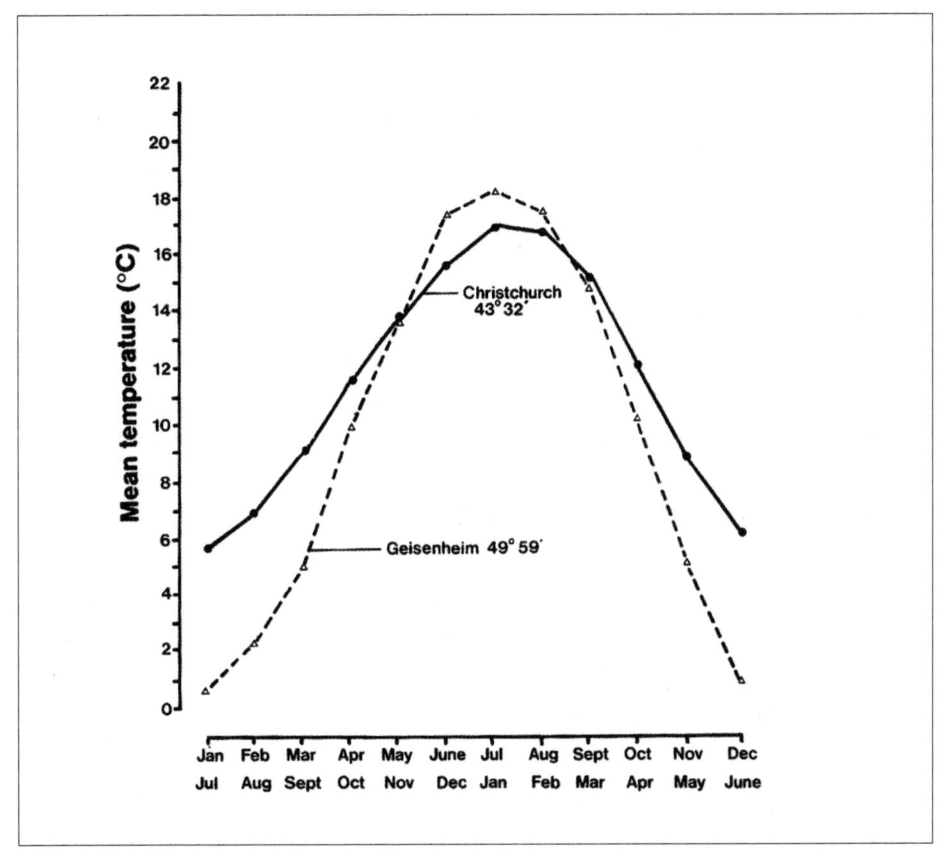

FIGURE 4.1 Heat distribution in two climates with similar Degree Days – Christchurch, New Zealand and Giesenheim, Rhine

does not grow Riesling, although close by on selected warm slopes this grape ripens well. Christchurch grows the same Riesling grape on the flat. Geisenheim Pinot noir is light of body, Christchurch Pinot is full-bodied. This and other information from the two areas suggests Christchurch has greater ripening capacity. The reason can be gleaned from Figure 4.1 which suggests that the extra heat that Christchurch accumulates by having a longer growing season is of more value than the hotter mid-summer at Geisenheim. Christchurch has a longer growing season primarily because it is closer to the equator. Geisenheim is more continental and this accounts for the hotter months in summer. It would not be helped by cutting off mid-summer temperatures at 19°C as suggested by Gladstones (1992) because in neither area is the MTWM above this figure. We have found that LTI is a very useful index for discriminating between cool climate areas, particularly those in Region I of the Amerine and Winkler DD classification. It is not appropriate for evaluating seasonal differences in heat accumulation.

THE USEFULNESS OF CLIMATIC INDICES

Seasonal comparisons

When comparing seasons for heat accumulation there is considerable value in using the concept of Degree Days (Heat Units). It does effectively discriminate between a warm and a cold year and allows one to monitor the heat accumulation during the season. An experienced grower will be able to predict harvest date with moderate accuracy by determining accumulated heat at different stages during growth. Thus if by, say, 31 June (31 December in the southern hemisphere) DDs are 150 more than normal then an early season could be predicted. Assessments of DDs closer to the end of the season give increasingly accurate predictions. Over the years growers will learn more about their site and will be able to relate DDs to quality parameters. Consequently, the grower will learn that to ripen say, Cabernet Sauvignon, adequately in his or her vineyard, a specific DD is needed. For example, it may be discovered that, for a particular district, heat accumulation below 1300 DDs gives an average wine with little character, 1300 – 1400 may produce good but not outstanding wines and above 1400 the wines will be of exceptional quality.

The grower will probably be able to relate his or her findings to those of neighbours in the same districts with similar soils and management practices. However, the more the vineyards vary the less accurate will be these comparisons.

The following indices have less or even no value for comparing seasons: Latitude; Mean Temperature of the Warmest Month; Length of Growing Season; Latitude Temperature Index.

Comparing districts

The big advantage of the Degree Day index is that it is useful for comparing seasons, but will also give a good approximation of the ability of a district to ripen a specific grape and even the style of wine that will be produced. As already indicated, DD is not always the most satisfactory index when one is comparing climates that differ in continentality and maritime factors, especially in cool climates. For these conditions our research indicates that LTI has considerable value. Below we compare the advantages and disadvantages of LTI and DD for evaluating climatic suitability.

TABLE 4.1 Comparisons of effectiveness of Latitude Temperature Index (LTI) and Degree Days (DD) for evaluating suitable districts for viticulture

DD	LTI
Separates moderately effectively all climates from cool to hot (Amerine and Winkler's Regions I to V).	Less appropriate for general separation.
Not especially reliable for cool climates.	Very effective for separating districts in Regions I and II of the Amerine and Winkler classification.
Do not indicate likelihood of winter damage in continental areas.	
Overestimates capacity of north-eastern USA and Canada.	
Both tend to overestimate ripening capacity of wet areas.	
	Tends to overestimate ripening capacity of high altitude areas.

Chapter 5 considers the practical value of such indices.

CHAPTER 5
Establishing a Vineyard – the Use of Climatic Understanding

A person wishing to establish a vineyard will normally have an idea of the area in which the vineyard should be placed. Maybe the potential grower already owns some land or he or she knows exactly the area where they wish to be situated. It might be because the area is already known to produce fine wines or it might just be an area where they wish to build a house and live a country lifestyle. Whatever the reason or desire it would be foolish to go ahead with vineyard establishment without a thorough investigation of all aspects of the site for potential grape/wine production.

This chapter investigates those aspects of development which pertain to climate. We look at:

1. Preliminary investigation of the district.
2. Using a climatic index to categorise the district.
3. Choosing appropriate varieties.

PRELIMINARY INVESTIGATION OF THE DISTRICT

The district for the vineyard may already be established and shown to be suitable for viticulture. The proposed site may be sufficiently similar to the surrounding vineyards that little further investigation may be required. Data for varieties to use and appropriate management to adopt may be obtained by observation and enquiry.

Often, however, there will be sufficient differences from the norm that these factors must be taken into account before a potential grower makes a decision.

Figure 2.8 shows variations that can be expected when aspect and slope differ. These variations must be considered because they will affect light interception heat accumulation and will suggest where frosts might be a problem.

CHOOSING AN INDEX

When an area has not previously been used for viticulture it would be wise for a grower to use a climatic index to help in assessment of likely advantages and disadvantages and the choice of suitable varietal mix.

The author has investigated a number of indices for their ability to predict the varieties that will grow in a specific location and found that in cool climates the Latitude Temperature Index (LTI) was the most accurate (Jackson and Cherry, 1988). We, therefore, recommend its use as an initial guide. Table 5.1 shows the LTI needs for several common varieties. We shall discuss this table in more detail shortly.

FIGURE 5.1 Stevenson Screen – right; Pan Evaporimeter, left. Photograph courtesy R A Crowder

GETTING THE DATA

Meteorological stations normally provide the data one needs for calculating LTI, Degree Days etc. In some cases the data from the station give an accurate indication of the values in the vineyard, but of course this is not always the case in different mesoclimates.

Growers can, if they wish, set up their own meteorological station using a Stevenson Screen. This is the world standard container in which are placed the thermometer(s) and other recording equipment – see Figure 5.1. Nowadays, many electronic devices will do the collection of data, but they must be calibrated against a Stevenson Screen so that the heat recorded is equivalent to that collected in a screen.

Other data recorded in a meteorological station and useful to the grower are rainfall and evapotranspiration. The latter records the loss of moisture from a surface of water in a Pan Evaporimeter (see Figure 5.1). This figure, expressed in millimetres per day, month, or year, is roughly equivalent to loss from a field completely covered with green leaf material, say a pasture. It tells us the amount of water loss so that, if required, we can monitor our irrigation. A vineyard which does not cover the whole area with leaves, does not lose quite that much water. As a rough approximation the water loss will be equivalent to the proportion of light not reaching the vineyard floor. If this is 50 per cent, then water needs will be approximately half of the measured evapotranspiration.

CHOOSING THE GRAPES FOR THE VINEYARD

Table 5.1 provides a list of grape varieties and the climates where they can be expected to grow – using LTI.

Amerine and Winkler's division of climates into Regions I, II, III, IV and V (page 46) is satisfactory for a broad classification but, as indicated earlier, DDs do not separate very well the cooler districts, which mostly fall into Region I. To retain the link with the earlier classification, LTI is used as an adjunct to separate districts in Region I. Thus the grouping: IA IB IC II III IV V. In fact most grape varieties will ripen once a climate is as warm as Region II. Those in Region II still make

TABLE 5.1 Grape varieties grouped according to ripening ability in different climates

Group and LTI		Grapes Grown
Region I		
Group IA: LTI <190	1. very cool	Siegerrebe, Ortega, Optima, Madelaine x Angevine 7672, Reichensteiner, Müller Thurgau, Seyval blanc, Huxelrebe, Bacchus.
	2. cool	Pinot gris, Pinot blanc, Pinot noir,* Pinot meunier,* Chasselas, Gewürztraminer, Sylvaner, Chardonnay,* Faberrebe, Kerner, Scheurebe, Auxerrois Aligoté.
Group IB: LTI 190–270	mild	The key varieties are Riesling and Pinot noir – the latter produces heavier, red wines, unlike the lighter wines made in Group IA. Chardonnay in such districts is more full-bodied.
Group IC: LTI 270–380	warm	The key varieties are Cabernet Franc, Merlot, Malbec, Sauvignon blanc, Sémillon and Chenin blanc. These are sometimes grown in the cooler Group IB, but seldom reach the same quality. Cabernet Sauvignon ripens in this group but seems to need LTI at the higher end of the range or in Region II to excel.
Region II: LTI 380 and above	warm – hot	Cabernet Sauvignon, Carignane, Grenache, Shiraz, Thompson's Seedless (Sultana), Cinsaut, Zinfandel. Once again, some of these will ripen in Group IC districts, but their cultivation is restricted mostly to warm to hot climates

*Those marked with an asterisk are especially suitable for producing Méthode Traditionelle (Champagne-style) wines.

good table wines with reds becoming dominant; Region III makes full-bodied red wines and Region IV makes fortified wines, while Region V is commonly used for table grapes and drying grapes.

Remember that warm climates will still produce wine from cool-climate grapes but the wines tend to lack finesse and elegance. This is, however, a generalisation, some varieties such as Chardonnay have a wide range of climatic areas where they produce good but different wine. This ranges from Group IA where it is used for Méthode Traditionelle–Méthode Champenoise through to IB where it produces full-bodied but delicate wine. Nevertheless, it is still able to produce distinctive wine in Group II.

Riesling and Cabernet Sauvignon are varieties which behave differently. Both appear to ripen at a similar rate and will be harvested at a similar time with similar sugar and acid levels; yet Riesling is properly a cool climate grape and produces the finest examples in those areas where the grapes ripen in the cooler autumn weather. If the area is too hot the wine loses its delicacy. Cabernet Sauvignon in Group IB will ripen but does not compete well with wines from Group IC. In fact, it is often its most reliable when grown in Region II from which fine full-bodied wines can still be produced.

In warmer climates, Gewürztraminer ripens at a similar time or even later than Pinot noir. This is because both varieties increase their sugar levels and lose acidity at a similar speed and ripeness occurs at the same date. In cool conditions, sugars rise at a similar rate but acid loss in Pinot is much slower than in Gewürztraminer; thus to gain adequate ripeness, for other than Méthode Traditionelle, Pinot noir requires more time, ie it ripens later.

Nothing is perfect and while climatic indices can be very useful for gauging the appropriateness of a site for specific grapes, other factors need to be considered. Here are some of them.

FACTORS THAT WILL DELAY RIPENING AND MIGHT MAKE A SITE HAVE LOWER RIPENING CAPACITY THAN THE INDEX PREDICTS (SEE ALSO TABLE 4.1)

1. High rainfall: if the growing season has two months or more of consistent rain or, where heavy rain is interspersed through the season, ripening may be delayed.
2. Proximity to a cold land mass (eg eastern USA and Canada) tends to make spring later and autumn sooner – resulting in lower than expected ripening capacity.
3. Areas with higher altitudes may have lower ripening capacity than predicted by LTI. The reason is that altitude lowers both the MTWM and the days available for growth; LTI picks up the first but not the second.
4. Mesoclimates may delay (and advance) fruit development (see Figure 2.8).
5. Excessive vegetation on the vineyard floor may delay development.
6. Heavy cropping can markedly delay maturity. The closer one is to the limits of the area to ripen a specific variety, the more care a grower needs to take so as not to overcrop.
7. Poor management of the vine's canopy can create excessive shading, leading to lower light near the bunches and cooler temperatures; therefore ripening of the berries is delayed.

CHAPTER 6
Cool Climates of the World

We have classified world climates according to various criteria, eg Degree Days, Latitude Temperature Index, etc. These indices indicate how warm a climate or a season is or how long a growing season can be. They enable us to predict which varieties to use in specific areas and possibly the degree of ripeness to be expected.

This chapter examines a few representative climates in France, Germany, USA, Canada, Australia and New Zealand.

SOUTHERN ENGLAND AND NOVA SCOTIA

Kew (London, UK) and the Annapolis Valley (Nova Scotia, Canada) are two areas on the northern limits of viticulture. Both concentrate on early-ripening varieties and both include a number of Franco-American hybrids in their varietal mix. On the other hand there are some major differences between the two areas. Southern England benefits from the Gulf Stream – warm ocean currents originating around the Caribbean that bring mild conditions to western Europe. Few other areas at this latitude (>50°) could contemplate viticulture. By contrast, the latitude at Nova Scotia (NS) is 44–45°. This is equivalent to Bordeaux, the Willamette Valley in Oregon, or Canterbury, New Zealand. We therefore need to ask why Nova Scotia is at the northern limits of viticulture while these other areas at similar latitudes can produce grapes in Groups IA, IB, or IC.

Lower-latitude districts usually have longer seasons but the Nova Scotia season (below 150 days) is short because of two major factors. First, a cold current (the Labrador Current) brings down cold water from the Arctic, cooling the north-east coast and, secondly, a massive cold land mass to the north and west provides a buffer of cold air which takes a long time to clear in the spring. Once summer does arrive, the temperatures rise to quite high levels (19.2°C) due to the continental factor and the relatively low latitude, which means the sun is relatively high in the sky.

At a cold –5°C in winter, injury to *Vitis vinifera* in Nova Scotia is common, whilst Britain, having a maritime climate, has a relatively mild MTCM of 4.2°C and few if any problems with winter injury. Rainfall at Kew is moderately low (594 mm) whilst at Nova Scotia it is high (1177 mm). Nova Scotia, therefore, is more likely to favour American or hybrid grapes which have greater resistance to cold and fungus.
The high rainfall in Nova Scotia will also tend to delay ripening.

Both English and Nova Scotia wines have won honours at wine shows but in world markets the cost/quality equation is not likely to be good. Thus it might be expected that both areas will rely on local pride and/or curiosity to sell the majority of their wines – unless global warming does in fact occur.

FIGURE 6.1 Kew (UK)

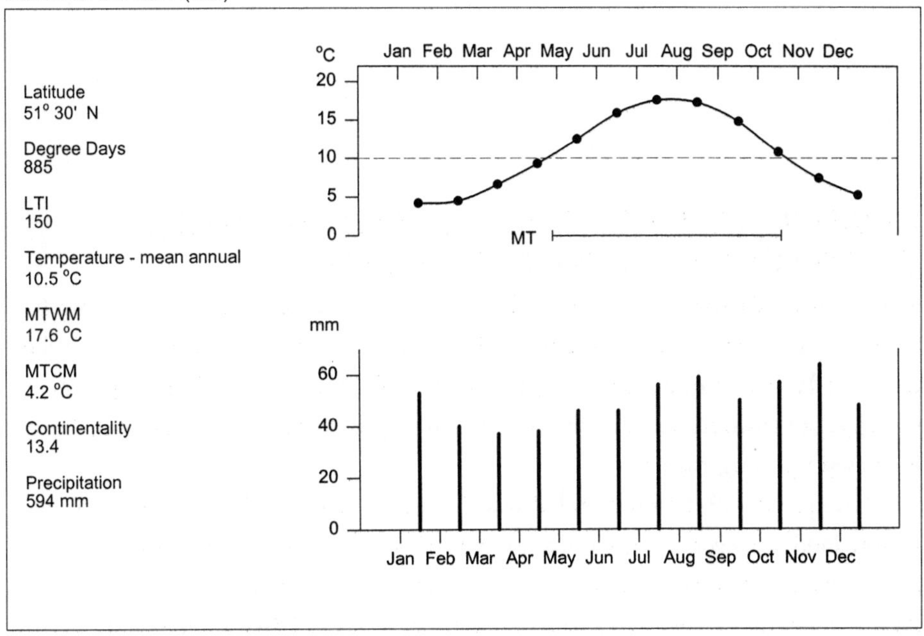

FIGURE 6.2 Kentville (Nova Scotia)

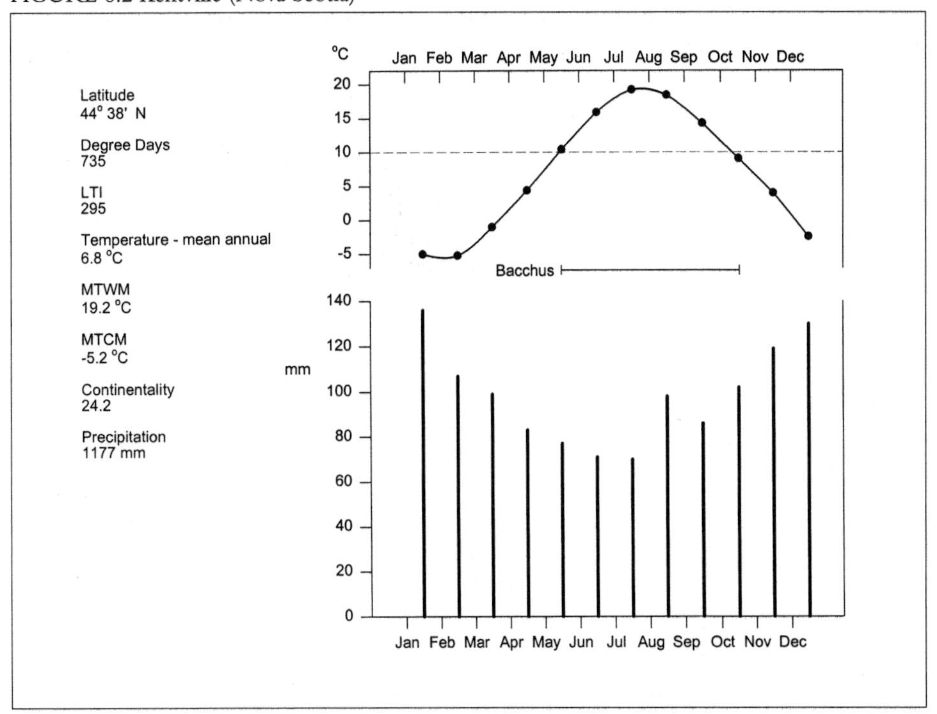

REIMS (CHAMPAGNE, NORTHERN FRANCE), CHRISTCHURCH (NEW ZEALAND)

Reims has similarities to the UK, such as equivalent Degree Days, rainfall, length of the growing season and continentality. LTI suggests that grapes will ripen slightly earlier at Reims than at Kew – which is probably true. In addition, grapes in the Champagne districts are usually planted on slopes so this may suggest a slightly increased ripening capacity.

Compared with Christchurch (New Zealand), Champagne has a shorter growing season (and a lower LTI), similar Degree Days and rainfall, but slightly higher continentality. Winter injury is unknown in Christchurch and not common in Reims. Pinot and Chardonnay ripen with more sugar in Christchurch and Riesling and Cabernet Sauvignon can be grown (although warmer mesoclimates are an advantage). Little 'Champagne' (Méthode Traditionelle) is grown in Christchurch but this is mostly for historical reasons.

FIGURE 6.3 Reims (Champagne district, France)

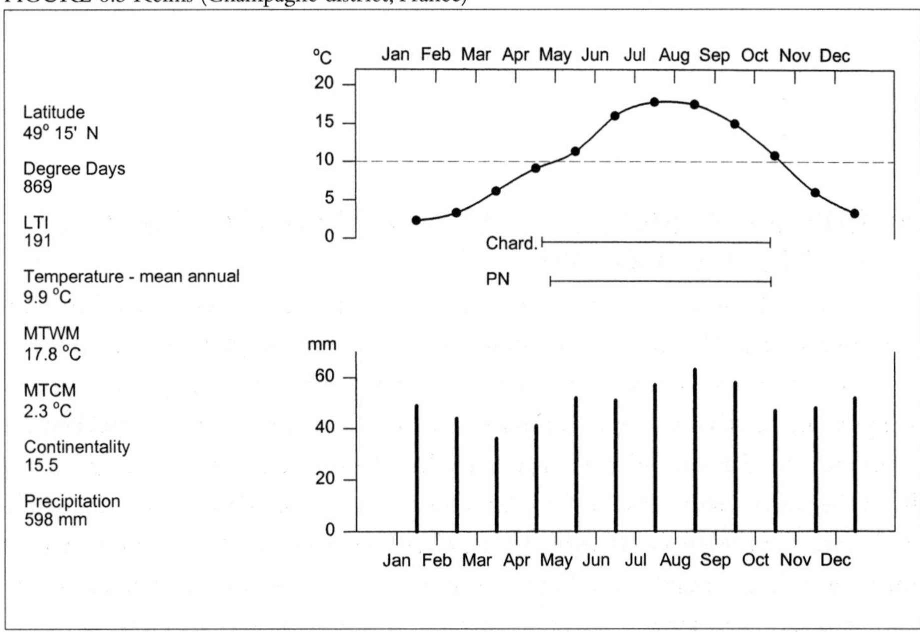

FIGURE 6.4 Christchurch (New Zealand)

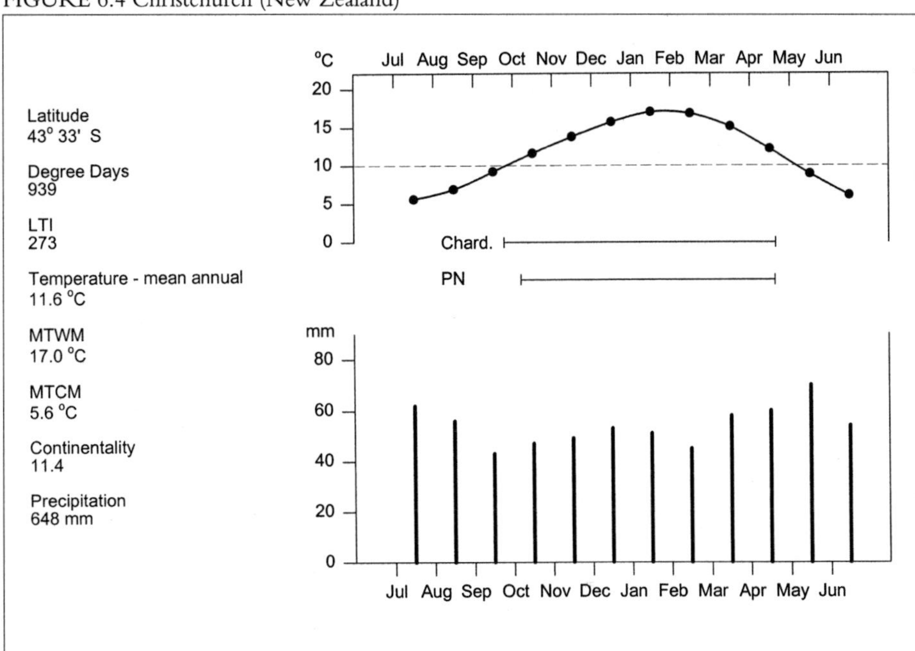

FREIBURG (GERMANY), SUMMERLAND (BRITISH COLUMBIA, CANADA)

As in Nova Scotia in the Canadian east, Summerland in the west has high continentality and loss from winter injury is not uncommon (MTCM can be as low as –5°C). Freiburg is less continental and suffers less damage from cold winter temperatures. Both have similar latitudes, Degree Days and LTI. Freiburg still benefits from the Gulf Stream and western Canada has slight amelioration of temperatures from the warm North-Pacific Current. Pinot noir and Riesling can be ripened in both areas. Summerland is very dry and has low disease pressure, Freiburg has moderately high rainfall but a big bonus is that while mid-summer rainfall is high there is a clear lowering of rainfall in autumn when the grapes are ripening.

FIGURE 6.5 Freiburg (Germany)

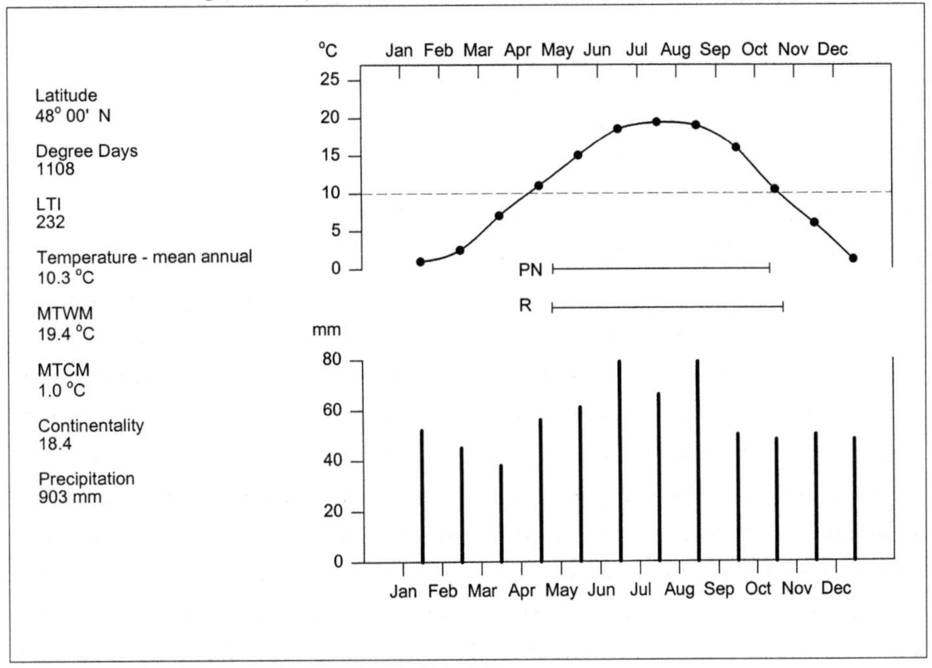

FIGURE 6.6 Summerland (British Columbia)

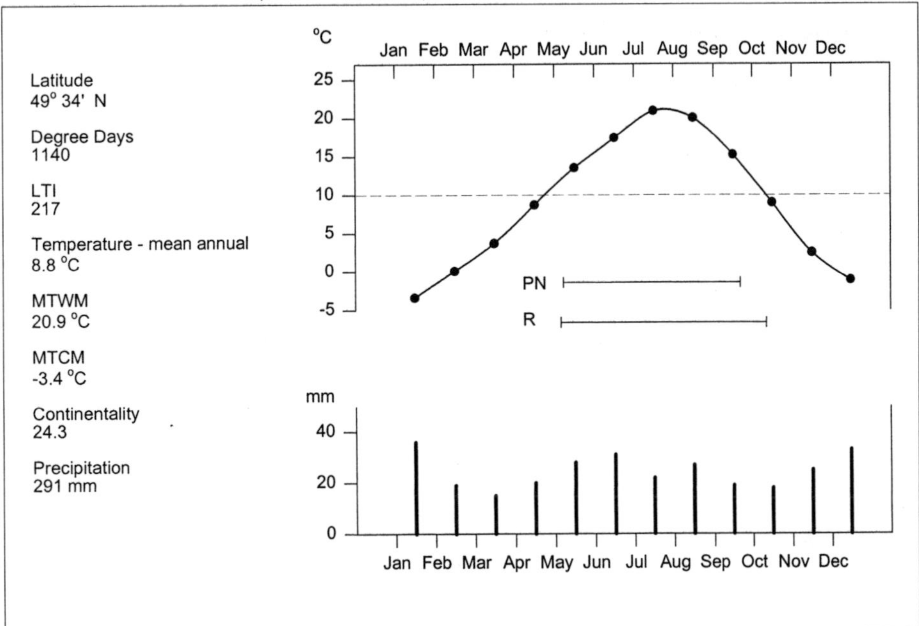

ZURICH (SWITZERLAND), CORVALLIS (WILLAMETTE VALLEY, OREGON)

These are two high-rainfall areas, both being above 1000 mm per annum, yet the pattern of distribution is very different. Zurich has the heaviest rain in mid-summer, although, fortunately for viticulture, rain becomes less heavy in the autumn period. Nevertheless, even here it is about 90 mm per month so we would expect botrytis and other wet-weather diseases to cause problems. Corvallis has almost no rain in mid-summer but it has wet springs and quite often wet autumns. Spring rain can delay bud burst and early winter rain means that growers tend to concentrate on earlier-ripening varieties. Consequently, both Corvallis and Zurich, due to high rain, do not grow some of the grapes that indices suggest they should – Zurich does not grow Riesling and the Willamette Valley has relatively few Cabernet Sauvignon. Overall, the conditions discussed suggest that as a grape-growing area the Willamette Valley has advantages over Zurich.

FIGURE 6.7 Zurich (Switzerland)

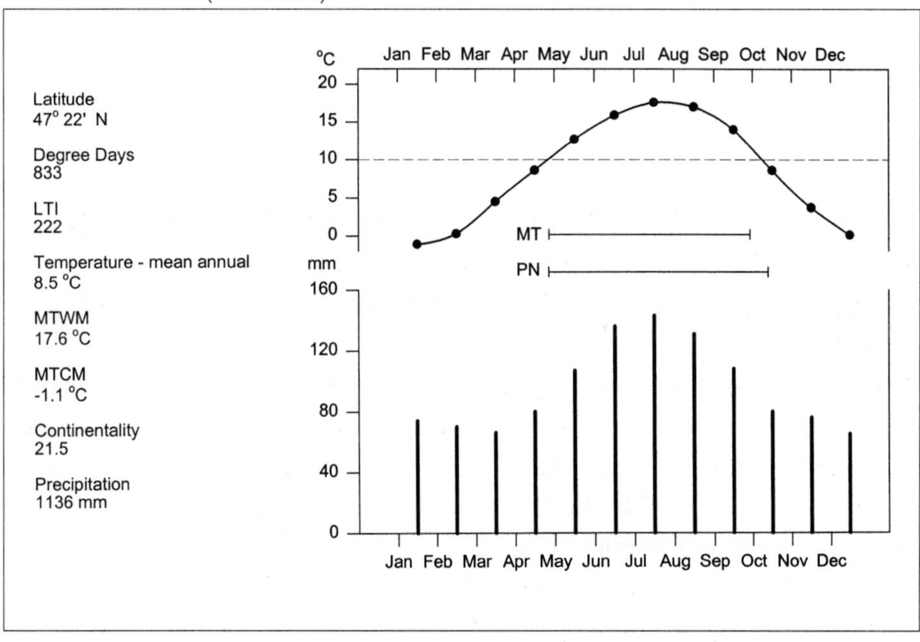

FIGURE 6.8 Corvallis (Willamette Valley, Oregon)

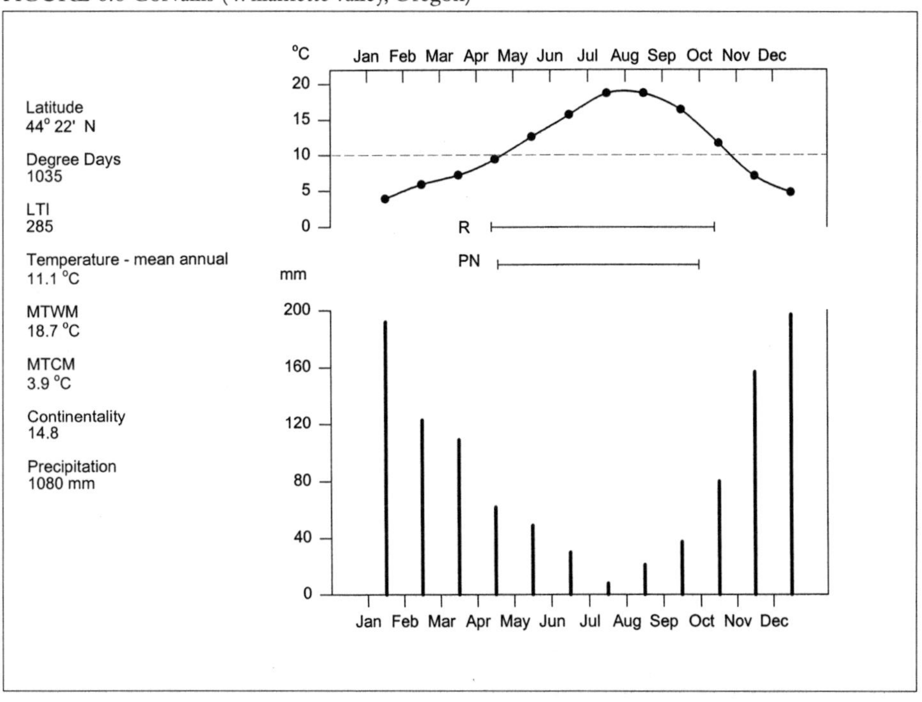

DIJON (BURGUNDY, FRANCE), CORVALLIS IN THE WILLAMETTE VALLEY (OREGON) AND CHRISTCHURCH (NEW ZEALAND) – THE PINOT FACTOR

The Burgundy area of France is renowned for its Pinot noir and Chardonnay. Pinot noir is of most interest here because until quite recently it has not been produced in such high quality elsewhere. Some New World areas are, however, now producing quality Pinot noir, notably the Willamette Valley in Oregon, New Zealand and some cooler parts of Australia.

LTIs are rather similar: 247 – Dijon; 285 – Corvallis; 273 – Christchurch.

All have reasonably dry ripening periods and are alpha zones for Pinot noir.

FIGURE 6.9 Dijon (Burgundy, France)

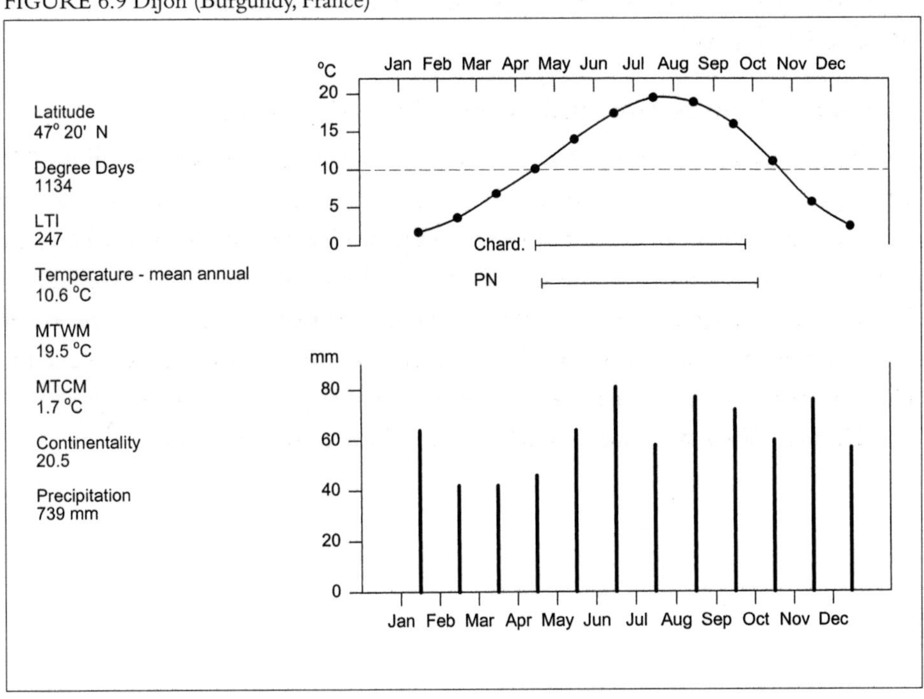

GRAPES IN THE COOL/WARM BOUNDARY OF VITICULTURE

Cabernet Sauvignon and other 'Claret' grapes

Cabernet Sauvignon will ripen in areas categorised as IC in this book (above 270 LTI). Nevertheless it also makes excellent wine in areas classified as II, above 380; although excessive heat is not conducive to the production of fine wines. Here we present data for areas producing Cabernet and related varieties. Most have LTIs of 290 to 425 or Degree Days of 1300 to 1600. Generally the temperature in the month before harvest is above 16°C and therefore these areas are beta zones and the climates are classified as warm rather than cool, but they tend to be marginal.

Data for Bordeaux, Napa, Hawkes Bay, and Mt Gambier are presented.

FIGURE 6.10 Bordeaux (France)

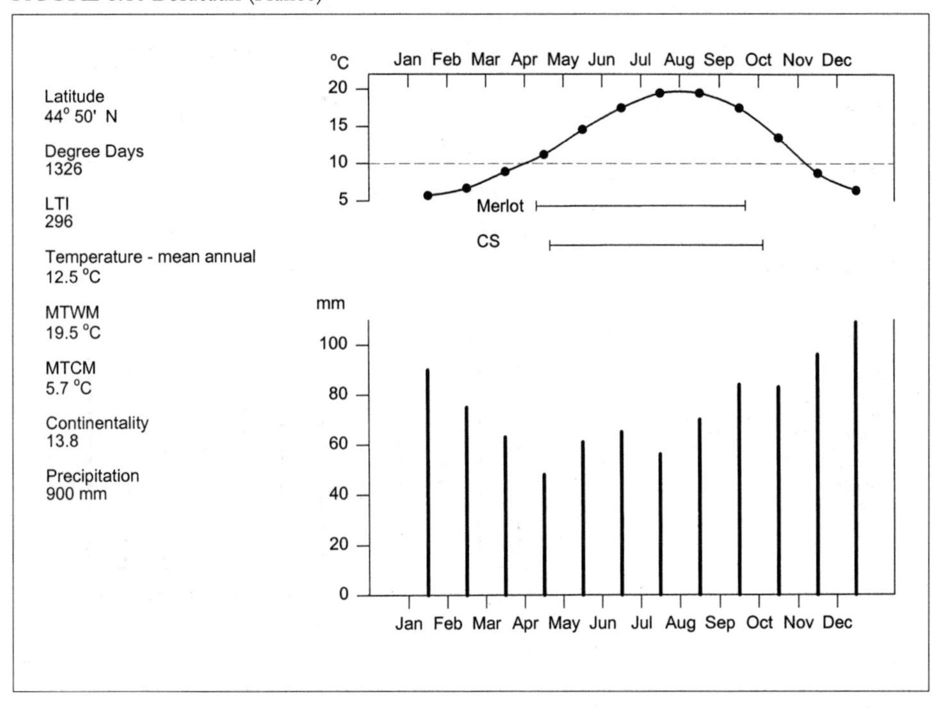

FIGURE 6.11 Napa (State Hospital, California, USA)

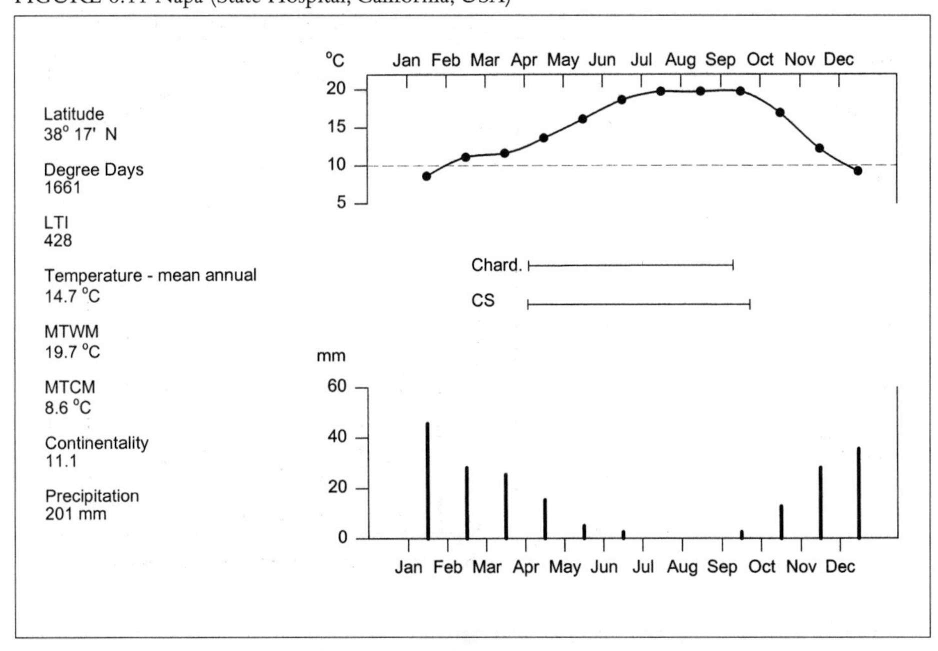

FIGURE 6.12 Napier (Hawkes Bay, New Zealand)

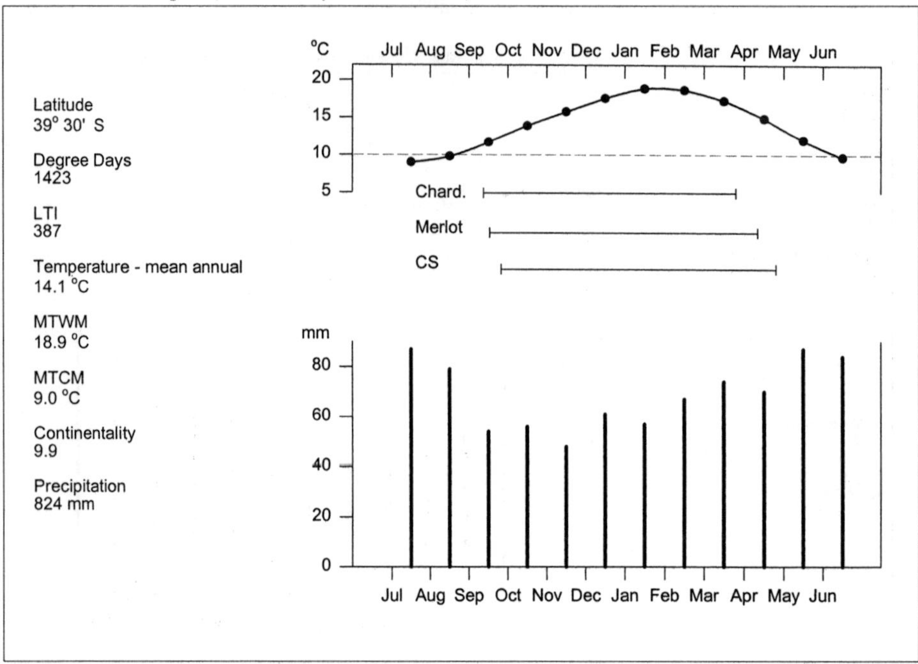

FIGURE 6.13 Mt Gambier (Australia)

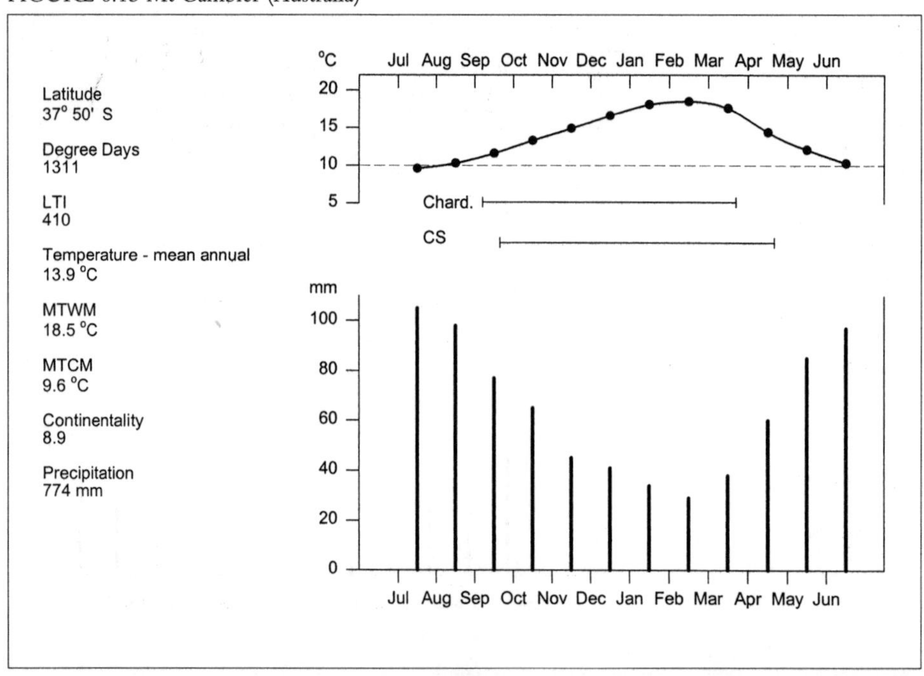

CHAPTER 7
Global Warming

Climates change and have been doing so since the world began. Previously this had been due to natural events, but we are now facing a situation where changes are occurring which may be due to human influence. In this chapter we will look at changes that have occurred historically, consider some of the evidence for the phenomena and examine some of the consequences of global warming – if it is occurring.

NATURAL CHANGES IN CLIMATE: SOME HISTORICAL CONSEQUENCES

When circumstances cause the earth to cool there are a number of responses, but, most spectacularly, ice sheets in the Arctic and Antarctic spread to lower latitudes. The increased ice cover accentuates the cooling because reflection of sunlight by the ice means less heat is absorbed by the Earth – so, once begun, an ice age may last for quite some time.

The last ice age ranged from about 12,000 to 65,000 years ago and ice sheets extended to middle-southern England, northern Germany, Canada, and northern USA. (The place where Chicago is now was under 1.5 km of ice.) The water locked up in the ice lowered the world sea levels, one consequence of which was to dry up the Bering Straits separating Siberia and northern America. This allowed humans, who had evolved in Africa and spread to Europe and Asia, to move into the American continent. Had viticulture been practised, 'cool-climate' wines would have been grown in southern Spain and northern Africa and warm-climate wines may have been produced in what is now the Sahara Desert.

The period from 10,000 to 12,000 years ago marked the beginning of the present *interglacial* and by 7000 to 5000 years ago a post-glacial optimum existed when temperatures were 2°C warmer than the present. Here are some of the subsequent changes that have occurred.

1. 5000 to 2500 years ago. The climate cooled and a 'little ice age' occurred.
2. 500 BC – 1300 AD. A warm period was experienced. Vikings colonised Iceland, Greenland and parts of North America. Romans brought the grape vine to England which became self-sufficient in wine. In the very warm period 1150 – 1300 AD in England vines were found 500 km north of the present limits.
3. 1300 – 1700 AD. Cooler conditions prevailed, the coolest being 1.5°C below those of the twentieth century. Poor harvests weakened the population and in the fourteenth century Black Death killed one-quarter of the population of Europe. The Vikings died out or abandoned Greenland. Ice fairs were held on the Thames.
4. 1700 – 1800 AD. Warmer conditions.
5. 1820 – 1830 AD. Very cold – the idea that Christmas should be white took hold in Britain.
6. 1850 – present day. Moderately warm conditions. Grapes re-introduced to England.

It is worth noting that while major ice ages affect the whole world, little ice ages and interglacials may vary between regions. For example, records suggest that the warm period around 1200 AD in Europe did not occur in China and Japan.

CAUSES OF CLIMATE CHANGE

Below I discuss how humans may be influencing change via the greenhouse effect. First, however, I consider some natural causes of climatic change.

Sun / earth cycles

The mathematician, Milankovicth, discovered cycles in the geometry of the sun-earth system that affect the climate irrespective of what happens within the earth. (For a discussion of this and other aspects see Gladstones, 1992.) There are three cycles that operate independently; first, a 100,000 year cycle due to changes in the earth's elliptical orbit, secondly, swings in the tilt of the earth's axis that occur over a 40,000 year cycle, thirdly, procession (wobbles) in the earth's axis of spin over 20,000 years.

Sunspot cycles

There are well known sunspot cycles of 10 to 12 year duration – peaks of sunspot activity are associated with warmer temperatures; low activity generally produces cooler conditions. There are also longer-term fluctuations, for example we are now in a period of higher than normal activity. The warm eleventh and twelfth centuries AD were also periods of prolonged sunspot activity, whilst the very cold period 1647–1715 was associated with very low activity.

El Niño / La Niña

These cycles are associated with an area of warm water of similar size to the continent of Europe moving from Indonesia to the South American coast and back (a distance of 4500 km).

In non-El Niño conditions, trade winds flow across the Pacific to the west and a surface layer of warm water piles up in the vicinity of Indonesia, raising the ocean surface 50 cm higher than in Ecuador. Cool nutrient-rich water wells up off western South America and this area supports an extensive fishing industry. During an El Niño period the easterly trade winds are lessened and warm water tends to move back to the east. This may result in intensified rainfall and flooding in Peru and droughts in Indonesia. Fishing in western South America can be hard hit and weather in other parts of the world may become unpredictable. Following El Niño there may be a La Niña which is a swing back to the opposite phase with cooler temperatures again in the seas off South America. Normally El Niño cycles are of 5–6 years duration but recently they have been shorter, making some observers think that they may be linked with global warming.

Cosmic collisions and volcanoes

Collisions can occur when a large meteorite or even a comet hits the earth. These, or large volcanoes erupting, cause dust, sulphuric acid droplets, and greenhouse gases (see later) to be released into the atmosphere. The dust and aerosol particles may move to and stay in the stratosphere for months or years and this can cause marked cooling by reflecting more sunlight away from the earth. One well-debated outcome of such action is the disappearance of the dinosaurs 65,000,000 years ago, which may have been due to cooling following a cosmic collision. The jury on this topic is still out.

The 1815 eruption of Mt Tambora in the Philippines probably caused the very cold temperatures in the second decade of the nineteenth century. Mt Pinatubo, also in the Philippines, erupted in 1991 and released massive amounts of dust, sulphur dioxide, and aerosols into the stratosphere. This was followed by three poor grape vintages in the southern areas of New Zealand.

HUMAN INPUT

The greenhouse effect

The greenhouse effect is so-called because certain gases in the atmosphere act like glass in a greenhouse. Most light wavelengths can pass through glass, and so heat the soil and other contents of the house. The warm soil loses heat, especially at night, but the heat is in the infra-red (long wavelength) form to which glass is more impervious. Thus the temperature of the glasshouse (or the earth) increases.

Humans are changing the level of some of these gases – with major effects on the earth's climate. Changes already established may cause an increase in world temperatures of 1–3°C per 100 years. If this were 3°C, then the sea level may rise 1 m due to the melting of ice and expansion of sea water as it warms up, with disastrous consequences for places like Bangladesh where 110 million people live on land barely above sea level.

The main 'greenhouse gases' are:

Water vapour: the dominant greenhouse gas which acts as a partial negative feedback because as warming occurs, more water vapour is taken up in the atmosphere, which increases cloud formation and reflects light away from the earth.

Carbon dioxide: Although CO_2 is less in volume than water vapour, the change in CO_2 is the main contributor to current warming. Our current use of fossil fuels such as coal, gas and oil, together with the burning of domestic fires and forests, plus the overall increase in industry that emits gases world wide, has increased the concentration of carbon dioxide in the atmosphere from 300 to 365 parts per million (ppm) over the last 100 years.

Methane: Intensification in agriculture and industry have increased methane concentration.

Ozone: is increasing partly as a result of industrialisation but less so than methane.

CFCs: (chlorofluorocarbons) are gases used in refrigerators and spray aerosols and will increase in concentration well into the twenty-first century when present cut-backs in use finally have their effects. (Note: CFCs and ozone are implicated in the so-called 'ozone hole'; this is a separate feature from their effects on global warming.)

Incidentally, increased carbon dioxide acts as an 'atmospheric fertiliser' and can promote increased growth of many plants. Providing that water and minerals are not limiting factors, more rapid growth can occur which in itself may remove more carbon dioxide, thereby counteracting global warming.

Global warming – some cautionary remarks

There is no doubt that the long-term effects of global warming on viticulture would be enormous, but we cannot discuss some likely consequences without a word of warning.

Most scientists agree that global warming will occur but the extent of the ensuing temperature change is much disputed. For example, doubts have been expressed that the temperature increases recorded in the last 100 years are indeed correct. Many meteorological stations have gradually become surrounded by urban development and it is known that urban areas may be 2°C warmer than rural areas. Thus, some of the recordings may be artifacts of urbanisation.

Gladstones (1992) has presented data from Ray (1968) which show that at Chateau Lafite in Bordeaux the date of picking has gradually become later over the 100 years until the late sixties suggesting cooling, not warming. In the 1860s the picking date was 16 September, in the 1950s and 1960s it was approximately 5 October. Some other factors such as the presence of new fungicides which allow later picking might modify the results but the data are hardly indicative of an increase in temperature in one of Europe's most prestigious vineyard areas. Similar work in Germany also shows the tendency towards later picking (Ray 1979).

Nature has a habit of making human efforts appear as naught and another serious volcanic eruption could even make us glad that some warming had occurred. Nevertheless, the 10 hottest years since 1860 have occurred in the 1980s and 1990s despite the eruption of Mt Pinatubo. At a recent meeting of the world's largest organisation of earth scientists, no speaker got up to argue that global warming was not occurring. Accordingly, it must be taken seriously.

CONSEQUENCES

We can predict some of the consequences of global warming for the grape–wine industry. Kenny and Shao (1992), using a modified Latitude Temperature Index, compared the likely northern limits of viticulture in Europe of a possible temperature rise of 1–4°C. An increase of 2°C due to global warming suggests the northern limit might include the Midlands and Norfolk in the UK, the Netherlands and northern Germany (see Figure 7.1). Areas such as the Rhine or Champagne could have ripening capacities similar to Burgundy or even Bordeaux. In fact in 1997 the British could have been forgiven for believing that global warming had already arrived – with summer temperatures 3°C above the long-term mean. Over the whole world the mean temperature in that year was 0.43°C above the long-term average.

So we have to ask the inevitable question: do we take global warming into consideration when we chose new areas for viticulture? Will Holland produce better 'Rhine' wines than the Rhine, will Oregon be the new Napa Valley and should our planting reflect these possibilities? I think that the taste and character we normally associate with specific wine areas will change – in some places for the better; in others not so. Still, we need to be cautious and it would be a mark of over-confidence to completely modify the varietal composition of a vineyard in the belief that a temperature rise of 1, 2, or 3°C was about to occur. We may however decide, if we are in a cool area, to be a little bold and plant a slightly higher proportion of the vineyard in later-ripening and currently marginal varieties.

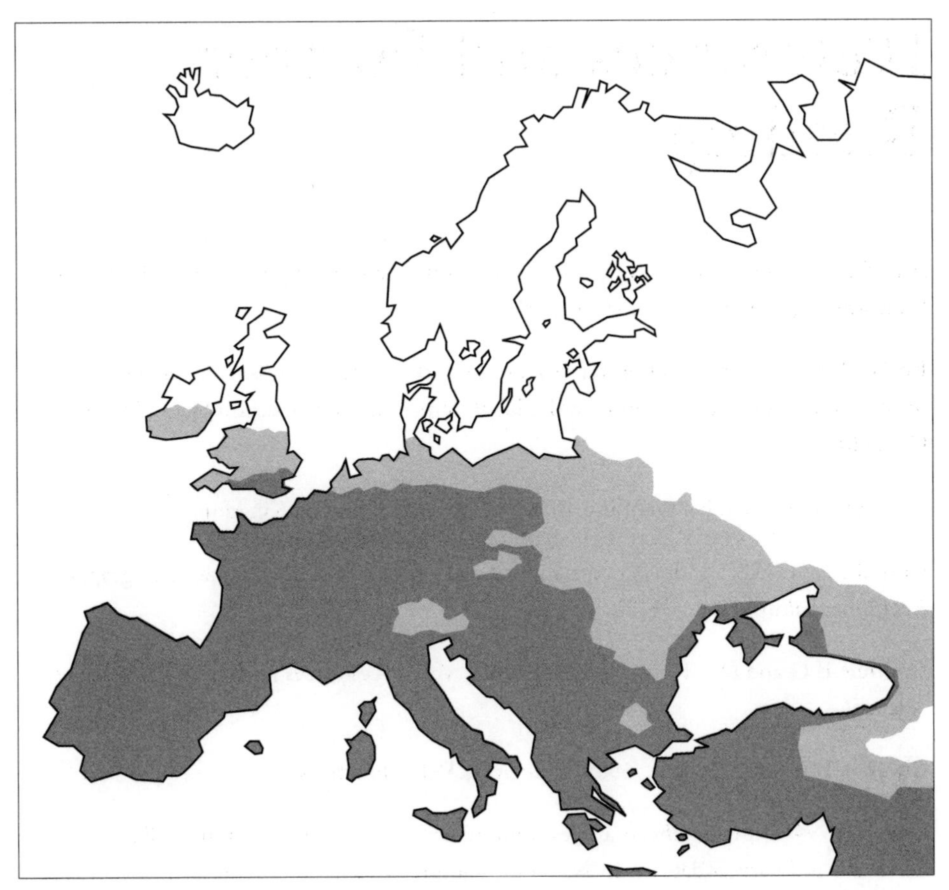

FIGURE 7.1 Europe showing current boundaries for viticulture (dark grey) and expected northern boundary if global warming raises temperatures by 3°C (light grey) – after Kenny and Shao (1992), who used a modified Latitude Temperature Index for these predictions

References and Further Reading

Amerine M A and Winkler A J (1944) Composition and quality of musts and wine of Californian grapes. Hilgardia. 15: 493–675

Becker N (1985) Site selection for viticulture in cooler climates using local climatic information. Proc. 1st Int Symp Cool Climate Vitic and Enol. Oregon State University. Corvallis.

Brown P (1996) Global Warming. Blandford, London, United Kingdom.

Coombe B G (1987) Influence of temperature on composition and quality of grapes. Acta Horticulturae. 206: 23–35.

Coombe B G and Dry P R (1988) Viticulture. Vol 1. Resources in Australia. Winetitles, Adelaide.

Coombe B G and Dry P R (1989) Viticulture. Vol 2. Practices.

Due G (1994) Climatic effects are less important than site and year in modelling flowering and harvest dates. Aust NZ Wine Industry Journal. 9: 56–66.

Due G (1995) Climate and wine: effects on quality and style. Aust NZ Wine Industry Journal. 10: 70–71.

Due G Pattison S Morris M and Coombe B G (1993) Modelling grapevine phenology against the weather consideration based on a large data set. Agriculture and Forest Meteorology. 5: 91–106.

Fullard H and Darby H C (Editors) (1979) New Zealand Concise Reference World Atlas. George Philip & Son, London.

Gladstones J (1992) Viticulture & Environment. Winetitles, Adelaide.

Gladstones J (1994) Climatology of cool viticultural environments: reply to criticism and a further examination of Tasmania and New Zealand. Aust NZ Wine Industry Journal. 9: 349–361.

Gladstones J Aust NZ Wine Industry Journal. 8: 317–318.

Gribben M and Gribbin J (1993) Being Human. Phoenix. London (contains a valuable discussion of the role of climate in human affairs).

Huglin P (1978) Nouveau mode d'evaluation des possibilities heliothermiques d'un milieu viticole. Acad Agric Fr CR Seances. 64: 1117–1126.

Jackson D I (1993) Climate debate: comments on 'Viticulture and Environment' by J Gladstones. 8: 317–318.

Jackson D I (1995) Climate – a tale of two districts. Aust NZ Wine Industry Journal. 10: 226–229.

Jackson D I (1995) Cool climate viticulture. Aust NZ Wine Industry Jourmal. 10: 362–365.

Jackson D I and Cherry N J (1988) Prediction of a district's grape ripening capacity using a latitude-temperature index (LTI). American Journal of Enology and Viticulture. 39: 19–28.

Jackson D and Looney N (1999) Temperate and Sub-Tropical Fruit Production.

Jackson D and Schuster D (1997) The Production of Grapes in Cool Climates. Lincoln University Press. Canterbury. New Zealand.

Jackson R S (1994) Wine Science – Principles and Application. Academic Press. San Diego.

Johnson J (1985) The World Atlas of Wine (3rd edition). Simon & Schuster. New York.

Kenny G J and Harrison P A (1992) The effects of climate variability and change on grape suitability in Europe. Journal of Wine Research. 3: 163–183.

Kenny G T and Shao J (1992) An assessment of a latitude-temperature index for predicting climate suitability for grapes in Europe. Journal of Horticultural Science. 67: 23–346.

Kirk J T O (1986) Application of a revised temperature system to Australian viticultural regions. Aust Grapegrower and Winemaker. 268: 48, 51–52.

Kirk J T O and Hutchinson M F (1994) Mapping of grapevine growing season temperature summation. Aust NZ Wine Industry Journal. 9: 247–251.

Martin D (date) pers comm.

Primault B (1969) Le climat et la viticulture. Int J Biometeorology. 13: 7–24.

Ray C (1968) Lafite. Peter Davies. London.

Ray C (1979) The Wines of Germany. Penguin. Harmondsworth.

Seguin G (1986) Terroirs and the pedology of wine growing. Experientia. 42: 861–873.

Wendland W M (1981) A fast method to calculate monthly degree days. American Meteorological Soc Bull. 64: 279–81.

Index

accumulated heat 16, 17
acetic acid bacteria 14
acid levels 2, 47
acid loss 47
acidic 4
acidity 47
acids 10
advective frosts 25
aerosol particles 66
Africa 63
air mixing 21
alcohol 4
alcohol level 2
Aligoté 46
alkathene tubing 23
alpha zone 5
alpha zones 5, 58
ambient temperatures 26
American 11, 50
American continent 63
Amerine 40, 42, 45
Amerine and Winkler classification 42
Annapolis Valley 49
Antarctic 63
Arctic 49, 63
aroma 4
Asia 63
Asian areas 9
aspect 44
atmosphere 66
atmospheric fertiliser 67
Australia 38, 49, 58
Australian 17, 36, 38
Austria 9
autumn frost 2
autumn frosts 10
Auxerrois 46

Bacchus 46
Bangladesh 66
bare soils 16
basal buds 22
Becker 4, 11, 26, 27
Bering Straits 63
berry death 2
berry splitting 30
beta zones 5, 60
Black Death 64
Blenheim 9
body 2, 3, 39
Bordeaux 9, 49, 60, 67, 68
Bordeaux districts 30
botrytis 14, 56
botrytis cinerea 14
'botrytised' wines 14
bouquet 4
Britain 50, 64
British 68
British Columbia 54
bud burst 1, 11, 21, 22, 36, 56
buds 2, 6, 7, 11, 30, 32
Burgundy 58, 68
Burgundy-style wines 3

Cabernet 11, 60

Cabernet Franc 46
Cabernet Sauvignon
3, 5, 40, 46, 47, 60
California 7, 9
Canada 9, 42, 48, 49, 54, 63
Canadian east 54
canes 6, 15, 22
Canterbury 49
capillary action 31
carbon dioxide 66
Caribbean 49
Carignane 46
Central Chile 7
Champagne 3, 33, 52, 68
Champagne districts 52
chaptalisation 3
Chardonnay 46, 47, 52, 58
Chasselas 46
Chateau Lafite 67
Chenin blanc 11, 46
Cherry 35, 44
Chicago 63
China 64
chlorofluorocarbons 67
Christchurch 39, 52, 58
Cinsaut 46
'Claret' grapes 60
clay soils 31, 32
climate index 38
climate indices 34, 35
climatic change 64
climatic index 43, 44
climatic indices 36, 40, 47
climatic zones 38
clone 29
clones 3
coal 67
coastal – maritime 9
cold 50
cold damage 11
cold temperatures 15, 66
comet 65
commercial wines 7
compounds 4
continental 8, 39, 54
continental areas 42
continental climates 8, 35
continental factor 50
continentality 8, 40, 52
cool 42, 46
cool climate 2, 3, 5
cool climate areas 39
cool climate districts 4, 37
cool climate grape 47
cool climates
3, 4, 9, 11, 16, 25, 40, 42, 44, 49
cool conditions 4, 5, 47
cool-climate grapes 47
cool-climate viticulturist 3
cool-climate wines 3, 4
cooler climates 2, 3
cooler regions 46
cooler temperatures 48
cooler zones 4
Coombe 4, 8
Corvallis 56, 58
cosmic collisions 65
crown gall infection 13
crushing 4
cycles 64

Darby 8
Degree Day index 42
Degree Days 34, 36, 38, 40, 42, 45,
49, 52, 54, 60
degree of ripeness 49
Degree-day accumulation 26
delicate wine 47
dessert grapes 18
dessert wines 4
devigoration 31
disease incidence 9

district 43
domestic fires 67
dormancy 7
dormant cuttings 32
dormant period 6
droughts 9
drying grapes 47
Due 36
Due, Graham 4

earlier-ripening 56
early-ripening 49
early-ripening grapes 38
earth's axis 65
easterly trades 65
eastern Australia 15
eastern USA 9, 48
Ecuador 65
elliptical orbit 65
emerging buds 11
England 38, 64
Europe 9, 63, 65, 68
European 11
Europe's 68
evapotranspiration 9, 45
excessive heat 60
excessive vegetation 48
extract 4

Faberrebe 46
fermentation 4
finesse 6, 47
flavour 4
flower bunches 6
fluctuations 65
forests 67
fortified wines 3, 9, 47
fossil fuels 66
France 49, 58
Franco-American hybrids 11, 49
freeze injury 12
Freiburg 54
Frost 11
frost 7, 11, 12, 15, 23, 25
frost control 11, 16, 18, 19, 21
frost damage 12, 16, 21
Frost reduction 22
frost risk 21, 25
frosts 6, 7, 11, 17, 18, 21, 22, 44
fruit 2, 10, 11, 13, 14, 15, 26, 30, 33
fruit crop 1
fruit development 31, 48
fruit quality 13
fruit splitting 2
full-bodied 39, 46, 47
Fullard 8
fungal diseases 14
fungal infection 33
fungicides 67
fungus 50

gas 67
gases 66
Geisenheim 39–40
German 14
Germany 2, 9, 14, 39, 49, 54, 68
Gewürztraminer 46, 47
Gladstones 36, 38, 40, 64, 67
global warming 50, 63, 65, 67, 68
graft union 15
grape grower 2
grape inflorescences 1
grape varieties 34, 45, 47
grape-growing area 56
grape-growing districts 39
grape-wine industry 68
Graz 9
greenhouse effect 64, 66
greenhouse gases 65, 66
greenhouse production 18
Greenland 64
Grenache 11, 46

growing season 52
Gulf Stream 38, 49, 54

hail 2, 6, 15
hail belts 15
hail damage 16
Hawkes Bay 5, 60
heat accumulation
2, 17, 25, 31, 34, 39, 40
heat application 18
heat units 34, 36
heavy cropping 48
high continentality 54
high latitudes 36
High Medoc 30
high rainfall 14, 35, 48, 50, 54
high temperatures 4, 10
high-rainfall areas 56
higher continentality 52
higher yields 2
Hobart 9
holistic 30
Holland 2, 68
hot 42
hot regions 4
Huglin 36
Hunter Valley 15
Huxelrebe 46
hybrid grapes 50

ice 23, 63, 66
ice age 63, 64
ice cover 63
ice sheets 63
Iceland 64
index 34, 35, 36, 37, 38, 39, 40, 44, 48
indices 34, 35, 41, 42, 44, 49, 56
Indonesia 65
inflorescence 1
inflorescences 6
infra-red 66
injury 15, 50
Inland 9
Inland – continental 9
integrated 30
interglacial 63
interventionalist 30
inversion strength 18
irrigation 8, 9, 17, 45
irrigation engineers 23

Jackson 3, 44
Japan 64
Johnson 8

Kenny 68, 69
Kerner 46
Kew 9, 49, 50, 52

La Niña 65
Labrador Current 49
latent heat 23
later ripeness 2
later-ripening 68
latitude 2, 35, 39, 40, 49, 50
Latitude Temperature Index
35, 37, 39, 40, 42, 44, 68, 69
latitudes 6, 8, 35, 49, 54
leaf fall 2
leaf variegation 12
Length of Growing Season 40
less damage 54
level of sugar 2
light frost 12
light interception 44
lighter wines 46
lignification 14
limestone soils 31
Lincoln University 39
Little 'Champagne' 52
little ice age 64
localised heating 18

London 9, 49
long-term average 34, 68
long-term mean 68
low acid levels 2
low disease 54
lower cropping 5
lower latitudes 36, 63
lower ripening capacity 48
lower sugar levels 3
lower-latitude districts 49
lowered ripeness 2

macroclimate 25
macroclimates 38
Madelaine x Angevine 7672 46
malate 4
malate dominance 11
malates 10
Malbec 45
management of vines 4
marginal varieties 68
maritime climate 50
maritime climates 9
maritime factors 40
mean monthly temperature 36
mean temperature 2, 5, 6, 36, 38, 68
mean temperature of the coldest month 11
mean temperature of the coldest month (MTCM) 7
mean temperature of the warmest month 38, 40
Mediterranean areas 7
Mediterranean Sea 7
Merlot 46
mesoclimates 5, 25, 29, 38, 45, 48, 52
meteorite 65
meteorological screens 36
meteorological stations 67
methane 67
Méthode Champenoise 47
Méthode Traditionelle 3, 47, 52
micro-jets 23
microclimate 14
microclimates 25, 26, 29
mid-summer temperatures 39
middle-southern England 63
Midlands 68
Milankovicth 64
mild 7, 36, 37, 49, 50
mild climates 3
Moselle 9, 14, 26, 33
Mt Gambier 60
Mt Pinatubo 66, 68
Mt Tambora 66
Müller Thurgau 5, 26
Müller Thurgau, 46

Napa 60
Napa Valley 68
Netherlands 68
New South Wales 15
New World areas 9, 58
New Zealand 5, 9, 39, 49, 52, 58, 66
nodes 30
Norfolk 68
North America 8, 64
North Island 5
North-Pacific Current 54
northern America 38, 63
northern Canada 38
northern Europe 42
northern France 9, 52
northern Germany 63, 68
northern hemisphere 37, 38
northern Italy 9
northern USA 63
Nova Scotia 49, 50, 54

oenology 4
oil 67
oil pot 20

oil pots 19
Ontario 9
open canopies 5
Optima 46
optimum rate of growth 10
Orange 15
Oregon 9, 49, 56, 58, 68
Ortega 45
oxidation faults 4
ozone 67
ozone hole 67

Pacific 65
Peru 65
pest build-up 8
Petite Syrah 3
pH levels 3
Philippines 66
photosynthates 36
photosynthesis 36
Pinot 39, 47, 52
Pinot blanc 46
Pinot gris 46
Pinot meunier 46
Pinot noir 3, 11, 39, 46, 47, 54, 58
Poland 9
Pomerol 31
poor grape vintages 66
Portland 9
post-glacial 63
primary factors 28
procession 65
Prosser 9
pruning 21, 29, 32, 36

radiation 27
radiation frosts 18
rain 56
Rainfall 50
rainfall 9, 17, 35, 45, 52, 54, *65*
rapid ripening 4
Ray 67
re-radiate 16
red wines 4, 46, 47
reds 3, 47
Reichensteiner 46
Reims 9, 52
resistance 11, 50
Rhine 9, 14, 25, 39, 68
Rhine valley 27
'Rhine' wines 68
Riesling 11, 26, 39, 46, 47, 52, 56
ripening ability 39, 46
ripening capacity
35, 37, 38, 39, 42, 48, 52, 68
ripening index 35, 36
ripening periods 58
Romans 64
root penetration 31
rootstocks 3
Russia 9

Sagan 68
Sahara Desert 63
San Francisco 9
Sauternes 14
Sauvignon 11
Sauvignon blanc 11, 46
Scheurebe 46
Schuster D 3
seasonal variation 34
secondary factors 28, 29
Seguin G 30
self-sheltering 17
Sémillon 11, 46
Seyval blanc 46
Shao 68, 69
shelter-belts 17, 25
Shiraz 3, 46
shoot growth 6, 12, 30, 31

shoot tipping 30
Siberia 63
Siegerrebe 46
site management 36
slope 25, 26, 29, 33, 44
'Smokeless' pots 18
Soil and plant management 29
soil management 29
soil temperature 32
South Africa 7
South America 65
South American 65
South Island 39
southern Australia 7
southern California 7
southern hemisphere 37, 38
southern Queensland 7
southern South Africa 7
southern Spain 63
specific areas 30, 49
split berries 13
splitting 33
spring frost 2
sprinkling rate 23
spur pruned shoots 6
St Emilion 31
Stevenson Screen 32, 45
stratosphere 66
style 4, 28, 29, 40
sub-tropical 1, 6, 7
sub-tropical areas 7
sub-tropical climates 7
sugar 3, 4, 10, 14, 47, 52
sugar levels 47
sugars 36, 47
sulphur dioxide 66
sulphuric acid droplets 65
Summerland 54
sun / earth cycles 64
sunburn 2, 11, 13
sunspot activity 65
sunspot cycles 65
Switzerland 9, 56
Sylvaner 26, 46

table grapes 47
table wines 3, 4, 8, 9, 47
tannins 13
tartrates 10
Tasmania 9
temperate climates 1, 6, 7
temperate crop 1
temperate regions 6
temperate zones 8
temperature 68
temperature change 67
temperature index 38
temperatures
5, 50, 54, 64, 65, 66, 68, 69
terroir 30, 32
the New World 9, 34
the Pinot factor 58
thermometer(s) 45
Thompson's Seedless 46
tolerance to temperatures 6
trade winds 65
training 21, 36
training methods 29
training systems 26
transpiration 14
trellis 15, 21, 32
trimming 30
Tropical 6
tropical areas 7
tropical climates 1, 6

varietal composition *68*
varietal mix 44, 49
véraison 14, 31
very cool 45

Vikings 64
vine canopy 15
vine damage 11
vine management 4, 21
vineyard floor 15, 45, 48
vintages 2
viticulturist 3, 18, 30, 36, 37
Vitis amurensis 11
Vitis vinifera 50
volcanoes 65

warm 45
warm climate 3, 5
warm climates 3, 4, 10, 47
warm regions 4
warm temperatures 6, 35, 36
warm-climate wines 3
warmer climates 2, 3, 47
warmer temperatures 65
Washington 9
water conservation 16
water loss 14, 17, 45
water penetration 31
water sprinkling 22
water stress 31
water vapour 66
Wendland 37
western Canada 54
western Europe 38, 49
western South America 65
wet areas 42
wet-weather diseases 56
white wines 4
whites 3
Willamette Valley 9, 49, 56
wind 6, 14, 15, 17, 25
wind breaks 17
wind machine 20
wind machines 21
wine style 29
Winkler 37, 39, 42
Winkler's 42, 45
winter damage 42
winter freeze 11, 15
winter freeze damage 13
winter frosts 11
winter injury 50, 52
world areas 9
world climates 9, 49

Zinfandel 46
Zurich 56